AF208169

Applied Mathematical Sciences | Volume 13

Rec'd May 5, '75

G. W. Bluman
J. D. Cole

Similarity Methods for Differential Equations

With 43 Illustrations

Springer-Verlag New York · Heidelberg · Berlin
1974

G. W. Bluman
Department of Mathematics
The University of British Columbia
Vancouver, Canada

J. D. Cole
Department of Mathematics
University of California
Los Angeles, California

AMS Classifications
Primary: 35A22, 35A25, 35A30, 34C20, 35K05
Secondary: 22#99, 33A75

Library of Congress Cataloging in Publication Data

Bluman, George W 1943-
Similarity methods for differential equations.

(Applied mathematical sciences; v. 13)
Bibliography: p.
Includes index.
1. Differential equations—Numerical solutions.
2. Differential equations, Partial—Numerical solutions.
3. Series, Lie. 4. Similarity transformations.
I. Cole, Julian D., joint author. II. Title.
III. Series.
QA1.A647 vol. 13 [QA372] 510′.8s [515′.35] 74-20838

Printed in the United States of America.

ISBN 0-387-90107-8 Springer-Verlag New York · Heidelberg · Berlin
ISBN 3-540-90107-8 Springer-Verlag Berlin · Heidelberg · New York

PREFACE

The aim of this book is to provide a systematic and practical account of methods of integration of ordinary and partial differential equations based on invariance under continuous (Lie) groups of transformations. The goal of these methods is the expression of a solution in terms of quadrature in the case of ordinary differential equations of first order and a reduction in order for higher order equations. For partial differential equations at least a reduction in the number of independent variables is sought and in favorable cases a reduction to ordinary differential equations with special solutions or quadrature.

In the last century, approximately one hundred years ago, Sophus Lie tried to construct a general integration theory, in the above sense, for ordinary differential equations. Following Abel's approach for algebraic equations he studied the invariance of ordinary differential equations under transformations. In particular, Lie introduced the study of continuous groups of transformations of ordinary differential equations, based on the infinitesimal properties of the group. In a sense the theory was completely successful. It was shown how for a first-order differential equation the knowledge of a group leads immediately to quadrature, and for a higher order equation (or system) to a reduction in order. In another sense this theory is somewhat disappointing in that for a first-order differential equation essentially no systematic way can be given for finding the groups or showing that they do not exist for a first-order differential equation.

Lie also investigated thoroughly first-order partial differential equations which are essentially equivalent to systems of first-order ordinary differential equations by the theory of characteristics. He also made a preliminary investigation of some second-order equations, for example, the heat equation, but did not develop the integration theory. During the last century these methods have been developed by various mathematicians, engineers, and physicists. A summary of the mathematical approach based on infinitesimal transformations is given in a recent Russian book, Group Properties of Differential Equations, by L. Ovsjannikov.

In the first part of this book the material on ordinary differential equations is reproduced in detail. A typical result is (1.14-2): the criterion that a first-order equation

$$\Omega(x,y,y') = 0$$

admit a given group, defined by infinitesimal generators $\{\xi(x,y),\ \eta(x,y)\}$ is that $\xi \frac{\partial\Omega}{\partial x} + \eta \frac{\partial\Omega}{\partial y} + \eta' \frac{\partial\Omega}{\partial(y')} = 0$ on $\Omega = 0$, for all (x,y) where

$$\eta' \equiv \frac{\partial\eta}{\partial x} + \left(\frac{\partial\eta}{\partial y} - \frac{\partial\xi}{\partial x}\right)y' - \frac{\partial\xi}{\partial y}\,y'^2 .$$

From a knowledge of the group the (general) solution can be expressed by quadrature.

In the second part of the book this method is extended to partial differential equations and several other connections are made. First the concept of the extended infinitesimal transformation is developed for several variables. It is next shown how invariance under a group can be used to reduce the number of independent variables. The resulting solutions are connected with the usual "similarity solutions" of partial differential equations. Since these solutions are sometimes obtained by physical dimensional analysis §2.5 discusses the connection between transformation theory and that method. A principal example which is treated here is the construction of the general similarity solution of the heat equation. Other examples, including some with non-linearities and some with boundary conditions are sketched but, of course, no complete catalog can be given.

In view of the pioneering work of Sophus Lie in pointing out the importance and use of infinitesimal transformations, the authors would respectfully like to dedicate this book to his memory.

The authors are indebted to F. Milinazzo for carefully reading drafts of the manuscript and suggesting numerous corrections and clarifications. Special thanks are due to Mrs. Vivian Davies and Mrs. Yit-Sin Choo for patiently deciphering the authors' handwriting in typing several drafts of the manuscript. The authors appreciate the technical assistance of the staff at Brown University.

George W. Bluman, Vancouver

and

Julian D. Cole, Los Angeles

TABLE OF CONTENTS

INTRODUCTION

The work presented here falls naturally into two parts; the aims and partial contents of these parts is now sketched for purposes of orientation.

1. Ordinary Differential Equations

For ordinary differential equations the aim is a theory of integration or reduction to quadratures. For first-order equations this means that special cases can be reduced to essentially the same case, one of quadrature. The canonical case of quadrature occurs when a variable is missing. If the general equation is

$$F\left(x, y, \frac{dy}{dx} \right) = 0 \tag{1}$$

the special case is

$$F\left(x, \frac{dy}{dx} \right) = 0. \tag{2}$$

For one of the (possibly many) branches of (2) we can write

$$\frac{dy}{dx} = G(x), \tag{3}$$

and the general solution is

$$y = \int_{x_0}^{x} G(\zeta)d\zeta + \alpha, \quad \alpha = \text{const.} \tag{4}$$

In the sense represented by (3) and (4) the problem is regarded as solved or reduced to quadrature. In this sense the theory accomplishes all that can reasonably be expected of it.

For higher order equations or systems the aim is the reduction of the problem to lower order plus a suitable number of quadratures, and this can be carried out for a definite class of problems.

2. Partial Differential Equations

For first-order partial differential equations we take the (restricted) point of view that a sufficiently complete integration theory is given by the theory of characteristics. This connects the solution of partial differential equations with the integration of systems of ordinary differential equations and hence with results of part 1. It may, however, be useful to look at some first order equations directly from the point of view of transformations and invariance.

For higher order equations or systems the aim is a reduction in the number of variables. A typical result is the statement that a solution $u(x,y,t)$ of a particular P.D.E. in three independent variables must be representable as

$$u(x,y,t) = \frac{1}{t} F\left(\frac{x}{t} , \frac{y}{t} \right). \qquad (5)$$

This procedure can possibly be repeated more than one time. The special case when a partial differential equation contains only two independent variables is particularly important since the problem is reduced to an ordinary differential equation. Further, the methods of part 1 may be applied. In many physical problems of interest the resulting equation which needs to be studied (together with a suitable number of quadratures) is of first order. In this favorable case the structure of all possible solutions in the phase plane provides complete information on the structure of a class of solutions to the original partial differential equation. It also may provide the basis for a method of numerical integration.

Another method, which can be used to obtain the same results in special cases, arises not directly from transformation theory but rather from dimensional analysis. The basic idea is that all physical problems must be expressible in dimensionless variables. This idea is applied to the variables entering a problem for a partial differential equation. For example if (x,y), some independent variables, which are space coordinates with the physical dimensions of "length", enter the problem, then it can be concluded that only the combination (x/y) (or equivalent) can enter the problem. Evidently, this represents a reduction in the number of variables. However, it should be remarked that the failure of dimensional analysis to predict similarity (i.e., a reduction in the number of essential variables) does not necessarily rule out similarity for the problem. The connection between dimensional analysis and similarity is discussed later.

What has been outlined above is the main content of this book. However, various related topics which enter naturally are discussed as the opportunity arises. Among these are asymptotic and local behavior, superposition of similarity solutions for linear cases.

Finally, it should be remarked that the methods used apply equally well to non-linear and linear cases. The ideas used represent one of the few systematic methods of attacking non-linear problems, with an eye to obtaining exact solutions.

1. ORDINARY DIFFERENTIAL EQUATIONS

1.0. Ordinary Differential Equations

The essential ideas of the method occur for first-order equations and these are discussed first. For first-order equations of first degree, which form the main subject matter of the first part of this book, the difference between the case when a variable is missing in the right hand side and the general case should be noted:

$$\frac{dy}{dt} = F(x,y) \qquad \text{general,} \tag{1.0-1}$$

$$\frac{dy}{dx} = F(x) \qquad y \text{ missing.} \tag{1.0-2}$$

In the general case the complete integration is represented by all the integral curves in the (x,y) plane, one curve passing through each nonsingular point (Fig. 1.0-1) according to the local direction field at each point P; these form ∞^1 (number of) curves.[1] Their construction demands in general ∞^1 integrations.

In the special case (1.0-2), again all the curves are needed for the complete solution. However, the complete solution, representing all the integral curves, is given indirectly by integration

$$y = \int_{x_o}^{x} F(\zeta)d\zeta + \alpha, \qquad \alpha = \text{const.} \tag{1.0-3}$$

Thus, essentially only one integration is needed; the problem is one of quadrature. This fact is reflected in the geometric properties of Fig. (1.0-2). The slope of each integral curve is the same at a fixed value of x. The integral curves are thus congruent and a

---

[1] Old fashioned notation: ∞^1 = single infinity of curves (characterized by all continuous values of one parameter).

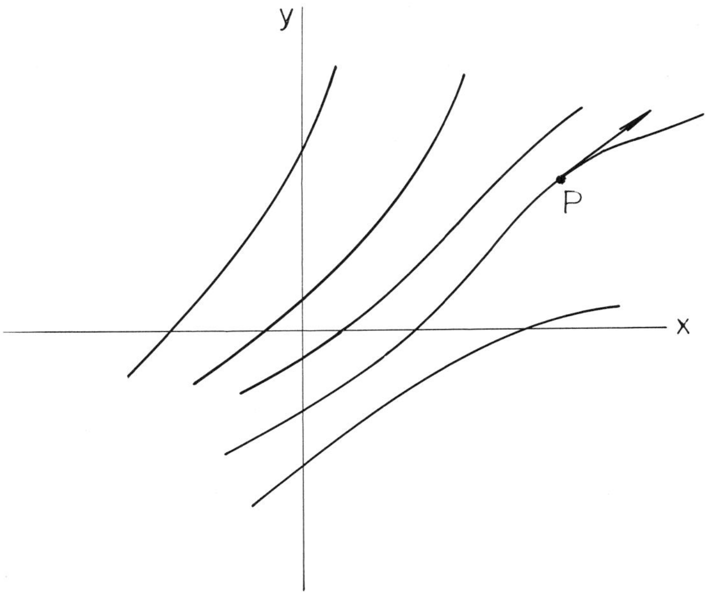

Figure 1.0-1

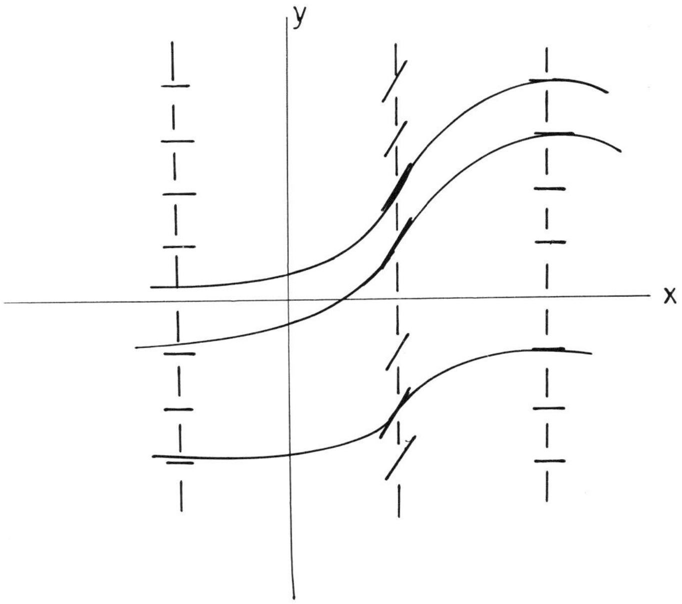

Figure 1.0-2

translation in the y-direction brings one into another. Thus, we can summarize the special properties of this case:

under the transformation $y \to y + \beta$:

- integral curves $\to$ integral curves (1.0-4)
- the differential equation (1.0-2) is invariant

The reduction to quadrature is the aim of the transformation theory for these first-order equations. According to the above remarks we might expect that invariance under transformation is the basic property which allows a reduction to quadrature. That this is so is illustrated by the special example of the next section. It is in fact possible to connect all transformations and invariances with that of (1.0-4).

1.1. Example: Global Similarity Transformation, Invariance and Reduction to Quadrature

This section demonstrates, in a special case, how invariance under a transformation can be used to reduce a problem to quadrature. Consider

$$\frac{dy}{dx} = F(x,y) \tag{1.1-1}$$

$F(x,y)$ is at first arbitrary but will soon be restricted by transformation requirements. Assume that the differential equation is invariant under the special transformation

$$\left.\begin{aligned} x^* &= \alpha x \\ y^* &= \beta y \end{aligned}\right\} \quad 0 < \alpha,\ \beta < \infty \tag{1.1-2}$$

(α,β)[1] are the parameters of the transformation. We consider the

---

(1) Greek letters will be used to denote parameters, as far as possible.

transformation of the original space (x,y) to an image space (x^*,y^*); this can also be thought of as the mapping of the plane into itself. The transformation assigns an image point $P^*(x^*,y^*)$ to each point $P(x,y)$ in the plane and vice-versa. The special transformation (1.1-2) is a stretching or similitudinous transformation. $\alpha = \beta = 1$ is the identity which is included in all transformations. A direction field at P^* is also assigned by the transformation of the differential equation

$$\frac{dy^*}{dx^*} = \frac{\beta}{\alpha} F\left(\frac{x^*}{\alpha} , \frac{y^*}{\beta} \right). \tag{1.1-3}$$

To the integral curve of (1.1-1) through P corresponds an integral curve of (1.1-3) through P^*.

Now we can define invariance precisely: <u>the differential equation (1.1-1) is said to be invariant under the transformation (1.1-2) when the differential equation reads the same in the new coordinates</u>. That is, the right hand side of (1.1-3) is equal to $F(x^*,y^*)$.

$$\frac{\beta}{\alpha} F\left(\frac{x^*}{\alpha} , \frac{y^*}{\beta} \right) = F(x^*,y^*) \quad \text{for invariance.} \tag{1.1-4}$$

We will assume that F is such that (1.1-1) is invariant.

Before considering the restrictions of $F(x,y)$ let us consider some other consequences of invariance. Consider a definite integral curve

$$y = f(x) \tag{1.1-5}$$

in the original space; that is, $f(x)$ is such that $f'(x) = F(x,f(x))$ for some range of x. The fact that the equation is invariant implies that the <u>same</u> curve in the (x^*,y^*) space is an integral curve of (1.1-3)

$$y^* = f(x^*)\,. \tag{1.1-6}$$

But each integral curve in the "star" space is the image of an integral curve in the original space. Upon transforming (1.1-6) back to the original (x,y) space we have, as integral curves (for various α,β)

$$y = \frac{1}{\beta}\, f(\alpha x)\,. \tag{1.1-7}$$

Thus, as a consequence of invariance, we can say that any integral curve in the original space such as (1.1-5) is a member of some family, such as (1.1-7). The identity member of this family has $\alpha = \beta = 1$. With the aid of (1.1-7) we can actually find the integral curve passing through any point of the plane. Thus, essentially only one integral curve needs to be calculated; the problem should be reducible to quadrature.

A procedure for doing this is now indicated. First, we find the form of $F(x,y)$. Two cases need to be considered:

(i) α,β independent parameters (the differential equation is invariant under a two parameter group); rewrite the invariance condition (1.1-4) as

$$\beta F(x,y) = \alpha F(\alpha x,\beta y)\,. \tag{1.1-8}$$

Then $\partial/\partial\alpha$ implies

$$0 = F(\alpha x,\beta y) + \alpha x \frac{\partial F}{\partial(1)}(\alpha x,\beta y)$$

or (1.1-9) [1]

$$0 = F(x^*,y^*) + x^* \frac{\partial F}{\partial x^*}(x^*,y^*)\,.$$

---

[1] The notation $\partial/\partial(1)$ means the partial derivative with respect to the first argument of the function; $\partial/\partial(2)$ denotes the partial derivative with respect to the second argument, etc.

Direct integration yields

$$F(x^*,y^*) = \frac{g(y^*)}{x^*} . \tag{1.1-10}$$

Thus, the basic functional equation (1.1-8) becomes

$$\beta g(y) = g(\beta y). \tag{1.1-11}$$

$\partial/\partial\beta$ of this functional equation yields

$$g(y) = yg'(\beta y) = \frac{1}{\beta} g(\beta y)$$

or (1.1-12)

$$\frac{g'(y^*)}{g(y^*)} = \frac{1}{y^*} .$$

The solution of (1.1-12) is

$$g(y^*) = by^* \qquad b = \text{const.} \tag{1.1-13}$$

and the resulting functional form of F is

$$F(x,y) = b\,\frac{y}{x} . \tag{1.1-14}$$

For this special differential equation

$$\frac{dy}{dx} = b\,\frac{y}{x}$$

a separation of variables provides the reduction to quadrature.

(ii) $\beta = \beta(\alpha)$ (the differential equation is invariant under a one parameter group).

The functional form of the dependence $\beta(\alpha)$ is not arbitrary but must be found in the course of finding the functional form of F. The basic functional equation (1.1-4) is now

$$\beta(\alpha)F(x,y) = \alpha F(\alpha x,\beta(\alpha)y) \tag{1.1-15}$$

$\partial/\partial\alpha$ implies

$$\beta'(\alpha)F(x,y) = F(\alpha x,\beta y) + \alpha x \frac{\partial F}{\partial(1)}(\alpha x,\beta y) + \alpha\beta' y \frac{\partial F}{\partial(2)}(\alpha x,\beta y) \quad \text{for all } (\alpha,\beta),$$

or the following first-order partial differential equation must be satisfied (when $F(x,y)$ is replaced from (1.1-15)):

$$\left(\frac{\beta'\alpha}{\beta} - 1\right)F(x^*,y^*) = x^*\frac{\partial F}{\partial x^*} + \beta'\frac{\alpha}{\beta}y^*\frac{\partial F}{\partial y^*}\ . \tag{1.1-16}$$

The characteristic differential equations[1] for (1.1-16) are

$$\frac{dx^*}{x^*} = \frac{dy^*}{\beta'\frac{\alpha}{\beta}y^*} = \frac{dF}{\left(\frac{\beta'\alpha}{\beta} - 1\right)F}\ . \tag{1.1-17}$$

Integration of the first two of these gives (α,β fixed) the curves $u(x^*,y^*) = \text{const.}$,

$$u = y^*/x^{*\frac{\beta'\alpha}{\beta}} \tag{1.1-18}$$

and integration of the first and third (along these curves) gives the general solution of (1.1-16).

$$F(x^*,y^*) = x^{*\left(\frac{\beta'\alpha}{\beta} - 1\right)} G\left(\frac{y^*}{x^{*\frac{\beta'\alpha}{\beta}}}\right) \tag{1.1-19}$$

where G is an arbitrary function. Now we can note that the original function $F(x^*,y^*)$ was by assumption free of any explicit dependence on the parameter α. Therefore

---

(1) See Appendix A for a discussion of the method of characteristics for first-order P.D.E.'s.

$$\frac{\beta'\alpha}{\beta} = \text{const.} = k. \tag{1.1-20}$$

Integration yields

$$\beta(\alpha) = \alpha^k \tag{1.1-21}$$

where the condition $\beta(1) = 1$ is used to identify the identity element of the transformation.

In summary, any differential equation of the form

$$\frac{dy}{dx} = x^{k-1} G\left(\frac{y}{x^k}\right) \tag{1.1-22}$$

is invariant under the one-parameter (α) family of transformations

$$\begin{aligned} x^* &= \alpha x \\ y^* &= \alpha^k y \end{aligned} \quad . \tag{1.1-23}$$

Note that

$$\frac{dy}{dx} = \frac{y}{x} H\left(\frac{y}{x^k}\right) \tag{1.1-24}$$

is a form equivalent to (1.1-22)

$$\left(G = \frac{y}{x^k} H\right).$$

Now the reduction to quadrature can be found by introducing the similarity coordinate $\sigma = y/x^k$ as a variable to replace x or y. σ is evidently invariant under transformation (1.1-23). In fact, this is a general rule: if the differential equation is expressed in terms of an invariant and any other appropriate coordinate, a reduction to quadrature is achieved.

In this case note that

$$\frac{d\sigma}{\sigma} = \frac{dy}{y} - k\frac{dx}{x} \tag{1.1-25}$$

while (1.1-24) is

$$\frac{dy}{y} = H(\sigma)\frac{dx}{x} . \tag{1.1-26}$$

Hence

$$\frac{d\sigma}{\sigma} = H(\sigma)\frac{dx}{x} - k\frac{dx}{x}$$

or

$$\frac{dx}{x} = \frac{d\sigma}{(H(\sigma) - k)\sigma} \tag{1.1-27}$$

(1.1-27) is the desired reduction to quadrature. Therefore

$$\log\frac{x}{x_o} = \int_{\sigma_o}^{\sigma} \frac{d\bar{\sigma}}{[H(\bar{\sigma}) - k]\bar{\sigma}} . \tag{1.1-28}$$

An alternative formulation expresses the invariance as a translation, similar to (1.0-4), with however more generality. The translation and corresponding congruence of integral curves takes place with respect to both variables. If we let

$$X = \log x, \qquad Y = \log y \tag{1.1-29}$$ (1)

then (1.1-24) becomes

$$\frac{dY}{dX} = H\left(e^{Y-kX}\right). \tag{1.1-30}$$

The invariance under the one-parameter family of transformations

$$X \to X + \gamma, \quad Y \to Y + k\gamma \tag{1.1-31}$$

---

(1) $x, y > 0$ with obvious changes if x or y is negative.

is now evident where $\gamma = \log \alpha$.

Problem 1.1-1. Consider the second-order differential equation

$$\frac{d^2y}{dx^2} = F\left(x, \frac{dy}{dx} \right) .$$

Assume that the differential equation is invariant under the special transformations

$$x^* = \alpha x$$
$$y^* = \alpha^k y .$$

Find the special form of the function F for which this is true. Show how the problem of obtaining the general solution is reduced to the integration of a first-order differential equation plus a quadrature.

1.2. Simple Examples of Groups of Transformations; Abstract Definition

It is clear from the previous work that a systematic study of transformations is a useful part of a general integration theory. As is discussed below these transformations must have group properties. In this section several simple examples of groups of transformations and associated concepts are introduced. In this and following sections we dispense with the "star" space and consider transformations of the plane into itself.

Translation Group in the Plane: One-Parameter Group

Consider the one-parameter family of translations which takes an arbitrary point (x,y) to another point (x_1,y_1) by a motion parallel to the y-axis (Fig. 1.2-1).

$$\left.\begin{aligned} x_1 &= x \\ y_1 &= y + \alpha \end{aligned}\right\} -\infty < \alpha < \infty \tag{1.2-1}$$

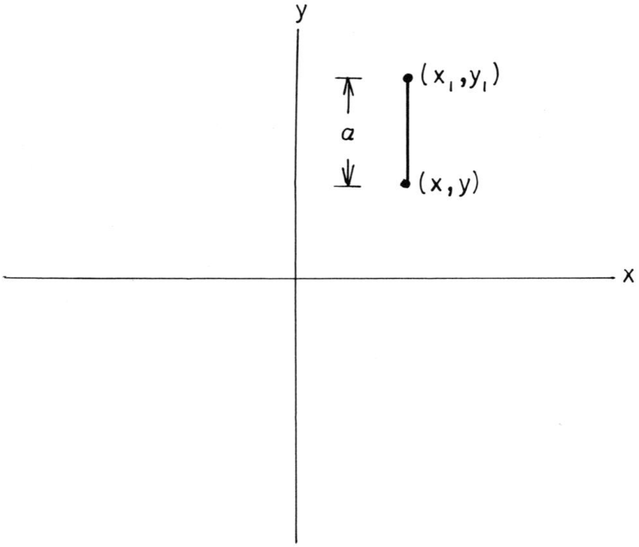

Figure 1.2-1

The transformation can now be repeated with a shift β to produce a new point (x_2, y_2)

$$\begin{aligned} x_2 &= x_1 \\ y_2 &= y_1 + \beta \end{aligned} \quad . \tag{1.2-2}$$

It is clear that the point (x_2, y_2) can be reached from the original point (x,y) by another transformation of the same family since

$$\begin{aligned} x_2 &= x_1 = x \\ y_2 &= y_1 + \beta = y + (\alpha+\beta) \end{aligned} \quad . \qquad (1.2\text{-}3)$$

The single transformation $(x,y) \to (x_2,y_2)$ has the parameter value $\lambda = \alpha + \beta$. The identity transformation ($\alpha = 0$ in (1.2-1)) is contained in the family as well as the inverse $(-\alpha)$.

Families of Transformations With These Properties are Said to Form a One-Parameter (α) Group

A more general abstract definition can be given to show when a family of transformations characterized by a continuous parameter (α), forms a group.

A one-parameter family (of transformations) in a set A (in this case a continuum of real values α) and a binary operation $\otimes$ in the set forms a group if the following axioms are satisfied:

Axiom 1 (Closure). For any elements α, β of A, $\alpha \otimes \beta$ is an element of A.

Axiom 2 (Associativity). For any elements (α, β, γ) of A

$$\alpha \otimes (\beta \otimes \gamma) = (\alpha \otimes \beta) \otimes \gamma .$$

Axiom 3 (Identity). There exists a unique identity element I in A such that for every element α of A

$$\alpha \otimes I = I \otimes \alpha = \alpha .$$

Axiom 4 (Inverse). For any element α of A there exists a unique α^{-1} of A such that

$$\alpha \otimes \alpha^{-1} = \alpha^{-1} \otimes \alpha = I .$$

For our purpose $\alpha \otimes \beta$ is an analytic function (γ) of α and β denoted by $\gamma(\alpha,\beta) = \alpha \otimes \beta$.

For the example of the family of translations in the plane given above the binary operation is addition of the shift distances, i.e., $\gamma(\alpha,\beta) = \alpha + \beta$; in other examples to be discussed later the binary operations may be multiplications or some other algebraic combinations. The associative property is the composition of shift distances and the identity and inverse are obvious.

Families of transformations may also form several parameter groups, for example translation in an arbitrary direction in the plane:

$$\left.\begin{aligned} x_1 &= x + \alpha \\ y_1 &= y + \beta \end{aligned}\right\} \text{parameters } (\alpha,\beta). \tag{1.2-4}$$

Since analytic dependence on the parameter (α) is always assumed quantities connected with differentiation with respect to the parameter play a central role in what follows. The simplest example of this is the infinitesimal transformation.

The <u>infinitesimal transformation</u> is arbitrarily close to the identity. Let α in (1.2-1) be an infinitesimal $\delta\tau$ then

$$\begin{aligned} x_1 &= x \\ y_1 &= y + \delta\tau \end{aligned} \tag{1.2-5}$$

and the original point is mapped in an infinitesimal neighborhood of the original point. The infinitesimal transformation can be repeated n times to give

$$x_n = x$$
$$y_n = y + n\delta\tau \qquad . \qquad (1.2\text{-}6)$$

According to the usual concept of integration, the global transformation (1.2-1) is produced in the limit $n \to \infty$, $\delta\tau \to 0$, $\Sigma n\delta\tau \to \alpha$.

Path curves are the curves traced out by a moving point (x,y) for a fixed initial point (x_0,y_0) as the parameter (α) assumes all

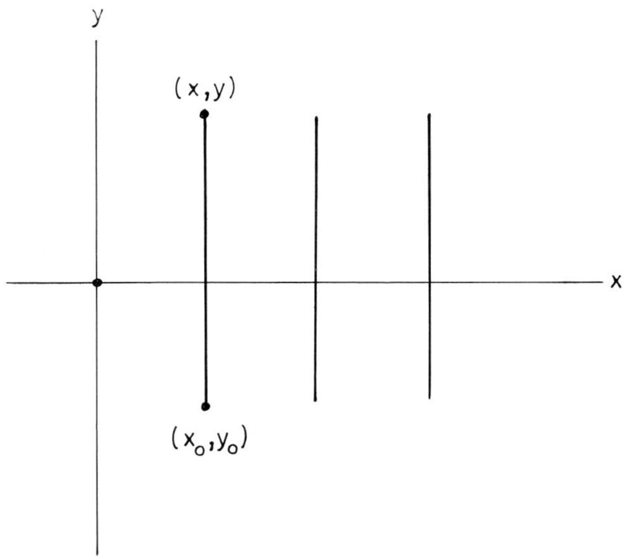

Figure 1.2-2

possible values $(-\infty,\infty)$. It is clear that the totality of path curves form a one-parameter family (e.g., the value of x_0 at $y_0 = 0$ identifies a path curve). The path curve as a whole goes into itself under any member of the transformation group (1.2-1) and is thus <u>invariant</u>. No other curves have this property in this case. Points at infinity can be discussed by projections.

Analytically, an <u>invariant function</u> over the plane $\Omega(x,y)$ (with respect to translation) is one whose value does not change under the transformation (1.2-1). That is

$$\Omega(x_1,y_1) \equiv \Omega(x,y+\alpha) = \Omega(x,y). \tag{1.2-7}$$

If we consider α infinitesimal the functional form of Ω is deduced analytically

$$\Omega(x,y+\alpha) = \Omega(x,y) + \alpha \frac{\partial\Omega}{\partial y}(x,y) + \frac{\alpha^2}{2!}\frac{\partial^2\Omega}{\partial y^2}(x,y) + \cdots = \Omega(x,y)$$

or, invariance implies

$$\frac{\partial\Omega}{\partial y} = 0, \qquad \frac{\partial^2\Omega}{\partial y^2} = 0 \quad \text{etc.}$$

and conversely. Hence

$$\Omega = \Omega(x). \tag{1.2-8}$$

Curves are represented by $\Omega(x,y) = \text{const.}$ and the <u>invariant</u> curves by $\Omega(x) = \text{const.}$ or $x = \text{const.}$, the path curves in this case.

<u>Example</u>: One-parameter group of rotations about the origin

$$\left.\begin{aligned} x_1 &= x\cos\nu - y\sin\nu \\ y_1 &= y\cos\nu + x\sin\nu \end{aligned}\right\}, \text{ parameter } \nu = \text{angle of rotation} \tag{1.2-9}$$

or

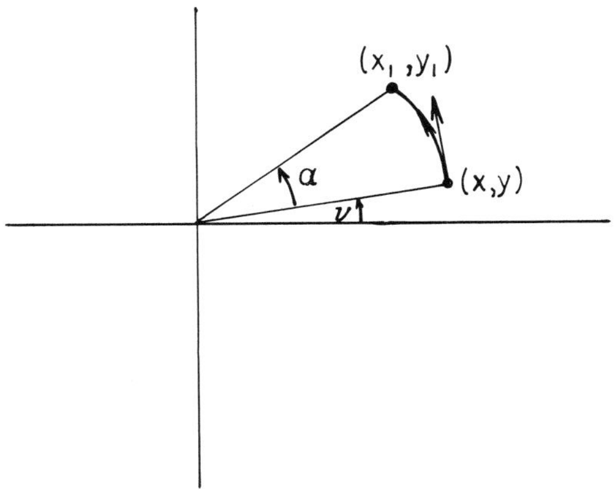

Figure 1.2-3

$$x_1 + iy_1 = (x+iy)e^{i\nu}. \qquad (1.2\text{-}10)$$

The identity element has $\nu = 0$, the inverse is $-\nu$, a rotation clockwise. The group property arises from the fact that rotation through ν followed by rotation through α is the same as rotation through $\gamma = \nu + \alpha$. The infinitesimal rotation is obtained from (1.2-9) with $\nu = \delta\tau \to 0$

$$\begin{aligned} x_1 &= x - y\delta\tau \\ y_1 &= y + x\delta\tau \end{aligned} \quad . \qquad (1.2\text{-}11)$$

The infinitesimal direction field of the path curves is given by

$$\begin{aligned} \delta x &= (x_1 - x) = -y\delta\tau \\ \delta y &= (y_1 - y) = x\delta\tau \end{aligned} \qquad (1.2\text{-}12)$$

This has the direction perpendicular to the radius $\delta y/\delta x = -x/y$. The path curves are evidently circles

$$x^2 + y^2 = x_o^2 + y_o^2 = \text{const.}$$

and thus each path curve is an <u>invariant</u> curve. Invariant functions for this group must satisfy

$$\Omega(x_1, y_1) \equiv \Omega(x \cos \nu - y \sin \nu,\ y \cos \nu + x \sin \nu) = \Omega(x,y) \quad (1.2\text{-}13)$$

or infinitesimally $(\nu \to 0)$, $\nu = \delta\tau$

$$\Omega(x - y\delta\tau,\ y + x\delta\tau) = \Omega(x,y).$$

Expanding,

$$\Omega(x,y) - \left[y \frac{\partial\Omega}{\partial x}(x,y) - x \frac{\partial\Omega}{\partial y}(x,y)\right]\delta\tau + \cdots = \Omega(x,y)$$

or invariance implies (and conversely)

$$y \frac{\partial\Omega}{\partial x} - x \frac{\partial\Omega}{\partial y} = 0. \qquad (1.2\text{-}14)$$

This first-order partial differential equation can be solved by characteristics; the general solution contains one arbitrary function ω and is

$$\Omega(x,y) = \omega(x^2+y^2). \qquad (1.2\text{-}15)$$

Invariance of the function corresponds to invariance of the path curves.

The same group of transformations can be expressed in other coordinates, in this case polar coordinates are convenient and (1.2-9) or (1.2-10) can be replaced by

$$\theta_1 = \theta + \nu, \quad r_1 = r. \tag{1.2-16}$$

In these canonical coordinates the rotation group is expressed as a translation in θ (1.2-1) and the transformation is said to be in canonical form.

Example: One-parameter group of affine transformations

The properties of this transformation are expressed in capsule form below

$$\left.\begin{aligned} x_1 &= \alpha x \\ y_1 &= y \end{aligned}\right\}, \quad 0 < \alpha < \infty \tag{1.2-17}$$

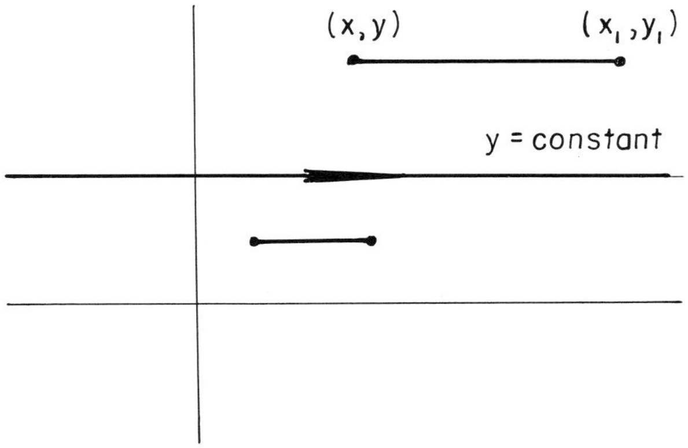

Figure 1.2-4 The transformation is a stretching in x proportional to x.

Summary of Properties:

parameter: α

identity: $\alpha = 1$

inverse: $1/\alpha$

composition: α followed by β yields a parameter value $\gamma = \alpha\beta$, i.e., $\gamma(\alpha,\beta) = \alpha\beta$.

infinitesimal transformation:

$$\alpha = 1 + \delta\tau \quad (x_1 = x + x\delta\tau,\ y_1 = y); \quad (\delta x = x\delta\tau,\ \delta y = 0).$$

path curves: $y = \text{const.}$

invariant curves: (1) $y = \text{const.}$

(2) $x = 0$; each point remains fixed.

invariant functions:

$$\Omega(x_1,y_1) \equiv \Omega(\alpha x,y) = \Omega(x+x\delta\tau,y)$$

$$= \Omega(x,y) + \delta\tau\left\{x\,\frac{\partial\Omega}{\partial x}(x,y)\right\} + \cdots = \Omega(x,y)$$

$$x\,\frac{\partial\Omega}{\partial x}(x,y) = 0$$

either $x = 0$ or $\Omega = \Omega(y)$; $\Omega(y) = \text{const.}$ on path curves.

canonical variables:

$$\left.\begin{aligned} r &= \log x \\ s &= y \end{aligned}\right\} \text{defined for } x > 0 \quad (\text{use } \log(-x) \text{ for } x < 0)$$

$$r_1 = \log x_1 = \log \alpha x = \log x + \log \alpha = r + \sigma$$

$$s_1 = y_1 = y = s$$

new parameter: $\sigma = \log \alpha$, $-\infty < \sigma < \infty$.

Thus, the use of canonical variables has changed a stretching group into a translation group.

1.3. One-Parameter Group in the Plane

The definitions introduced in §1.2 are generalized in this section to arbitrary transformations.

A one-parameter (α) family of transformations of the plane into itself

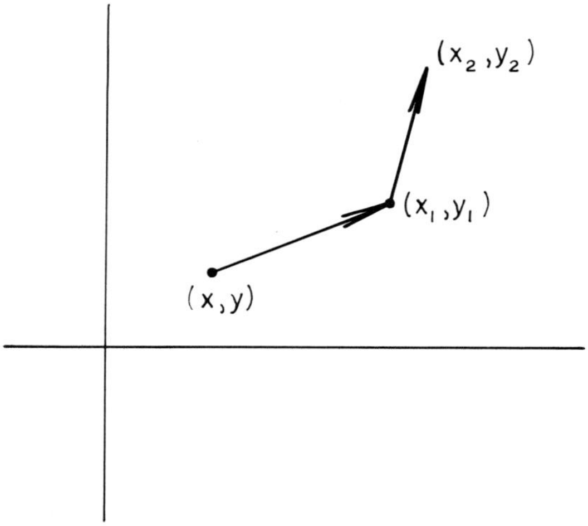

Figure 1.3-1

$$x_1 = \phi(x,y;\ \alpha)$$
$$y_1 = \psi(x,y;\ \alpha) \tag{1.3-1}$$

forms a group when to each (x,y) in the plane there corresponds one (x_1,y_1) and vice-versa, and when the group composition property holds.

The composition property demands that a repeated application of (1.3-1) can be expressed as a member of the same one-parameter family of transformations; that is a suitable parameter value can be found. Formally, we need

$$x_2 = \phi(x_1,y_1;\ \beta) = \phi(\phi(x,y;\ \alpha),\ \psi(x,y;\ \alpha);\ \beta) = \phi(x,y;\ \gamma)$$
$$y_2 = \psi(x_1,y_1;\ \beta) = \psi(\phi(x,y;\ \alpha),\ \psi(x,y;\ \alpha);\ \beta) = \psi(x,y;\ \gamma) \tag{1.3-2}$$

where

$$\gamma = \gamma(\alpha,\beta)$$

defines the law of composition, the binary operation in the set. Equations (1.3-2) must hold for all $(x,y;\ \alpha,\beta)$.

The assumed existence of the inverse element α^{-1} for which

$$\begin{aligned} x &= \phi(x_1,y_1;\ \alpha^{-1}) \\ y &= \psi(x_1,y_1;\ \alpha^{-1}) \end{aligned} \tag{1.3-3}$$

guarantees the existence of the identity element α_0; the transformation α followed by α^{-1} brings (x,y) back to (x,y)

$$\begin{aligned} x = \phi(x_1,y_1;\ \alpha^{-1}) &= \phi(\phi(x,y;\ \alpha),\ \psi(x,y;\ \alpha);\ \alpha^{-1}) \\ &= \phi(x,y;\ \alpha_0) = x,\ \gamma(\alpha,\alpha^{-1}) = \alpha_0 \\ y = \psi(x_1,y_1;\ \alpha^{-1}) &\quad \text{etc.} \end{aligned}$$

Note that not every one-parameter family of one-to-one transformations forms a group. For example, consider

$$\begin{aligned} x_1 &= \alpha - x \\ y_1 &= y \end{aligned}$$

a transformation with no identity element or composition law.

Once again we remark that the local structure of the transformation group is most important.

The analytic dependence[1] of (1.3-1) on α in the neighborhood of the identity element α_0

---

[1] The existence of a suitable number of derivatives of ϕ,ψ is assumed.

$$\left.\begin{aligned} x &= \phi(x,y;\ \alpha_o) \\ y &= \psi(x,y;\ \alpha_o) \end{aligned}\right\} \text{identity} \tag{1.3-4}$$

implies the existence of the infinitesimal transformations: Let $\alpha = \alpha_o + \delta\alpha$. Then (1.3-1) reads

$$\begin{aligned} x_1 &= \phi(x,y;\ \alpha_o+\delta\alpha) = \phi(x,y;\ \alpha_o) + \delta\alpha \frac{\partial\phi}{\partial\alpha}(x,y;\ \alpha_o) \\ &\qquad + \frac{(\delta\alpha)^2}{2!} \frac{\partial^2\phi}{\partial\alpha^2}(x,y;\ \alpha_o) + \cdots \\ y_1 &= \psi(x,y;\ \alpha_o+\delta\alpha) = \psi(x,y;\ \alpha_o) + \delta\alpha \frac{\partial\psi}{\partial\alpha}(x,y;\ \alpha_o) + \cdots \end{aligned} \tag{1.3-5}$$

or neglecting higher order terms, the infinitesimal transformation is

$$\begin{aligned} \delta x &= x_1 - x = \frac{\partial\phi}{\partial\alpha}(x,y;\ \alpha_o)\,\delta\alpha \\ \delta y &= y_1 - y = \frac{\partial\psi}{\partial\alpha}(x,y;\ \alpha_o)\,\delta\alpha \end{aligned} \quad . \tag{1.3-6}$$

It will be shown in §1.4 that for a given parameterization it is impossible that both $\frac{\partial\phi}{\partial\alpha}(\alpha_o),\ \frac{\partial\psi}{\partial\alpha}(\alpha_o) \equiv 0$. Let $\xi(x,y) = \frac{\partial\phi}{\partial\alpha}(x,y;\ \alpha_o)$, $\eta(x,y) = \frac{\partial\psi}{\partial\alpha}(x,y;\ \alpha_o)$. Then (1.3-6) can be written in the form:

$$\begin{aligned} \delta x &= x_1 - x = \xi(x,y)\,\delta\alpha \\ \delta y &= y_1 - y = \eta(x,y)\,\delta\alpha \end{aligned} \tag{1.3-7}$$

(ξ,η) depend on the original form (1.3-1) so that α_o is fixed and need not be shown in (ξ,η). The functions (ξ,η) define the transformation locally and we now show that in fact the global

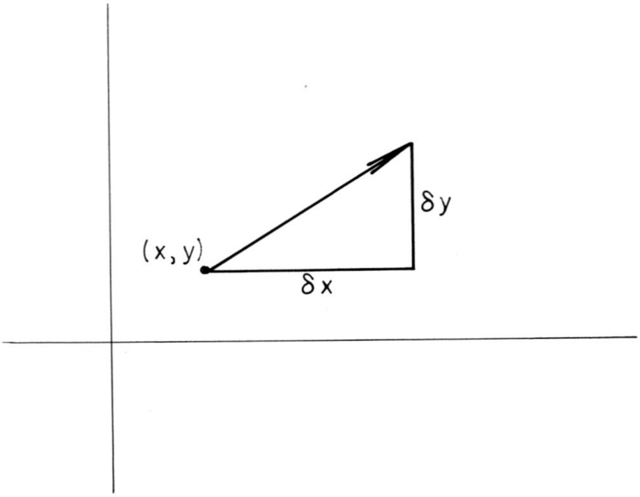

Figure 1.3-2

transformation can be reconstructed from (ξ,η). That is, the local transformation contains all the essential information about the global group of transformations.

Construction of the Group from ξ,η

A given infinitesimal transformation (ξ,η) attaches an infinitesimal direction field to each point (x_1,y_1) of the plane such that

$$\frac{dx_1}{\xi(x_1,y_1)} = \frac{dy_1}{\eta(x_1,y_1)} = d\tau. \qquad (1.3\text{-}8)$$

As the parameter τ varies all points of the plane undergo a motion;

repeated application of (1.3-8), equivalent to the usual process of integration, generates a curve from each original point. These curves are represented by the global transformation equations and we now show that they form a group. First we consider local behavior based on (1.3-8).

Along a curve $x_1(\tau)$, $y_1(\tau)$ the higher derivatives corresponding to (1.3-8) are:

$$\frac{dx_1}{d\tau} = \xi(x_1,y_1), \qquad \frac{dy_1}{d\tau} = \eta(x_1,y_1)$$

$$\frac{d^2x_1}{d\tau^2} = \frac{\partial\xi}{\partial x}(x_1,y_1)\frac{dx_1}{d\tau} + \frac{\partial\xi}{\partial y}(x_1,y_1)\frac{dy_1}{d\tau} = (\xi\xi_x+\eta\xi_y) \quad \text{at} \quad (x_1,y_1)$$

$$\frac{d^2y_1}{d\tau^2} = \frac{\partial\eta}{\partial x}\frac{dx_1}{d\tau} + \frac{\partial\eta}{\partial y}\frac{dy_1}{d\tau} = (\xi\eta_x+\eta\eta_y) \quad \text{at} \quad (x_1,y_1).$$

Therefore, for x_1,y_1 close to the initial point (x,y), $(\tau \to 0)$ we have

$$\begin{aligned} x_1 &= x + \tau\xi(x,y) + \frac{\tau^2}{2}(\xi\xi_x+\eta\xi_y) + \cdots \\ y_1 &= y + \tau\eta(x,y) + \frac{\tau^2}{2}(\xi\eta_x+\eta\eta_y) + \cdots \end{aligned} \tag{1.3-9}$$

This power series will, in general, converge in some neighborhood of $\tau = 0$ and represents the global transformation in the neighborhood of convergence.

The group composition property of the power series representation (1.3-9) is easily demonstrated. Proceeding from x_1 to x_2 with a parameter β (along the same curve)

$$\begin{aligned} x_2 &= x_1 + \beta\xi(x_1,y_1) + \frac{\beta^2}{2}\left[\xi(x_1,y_1)\xi_x(x_1,y_1) + \eta(x_1,y_1)\xi_y(x_1,y_1)\right] + \cdots \\ y_2 &= y_1 + \beta\eta(x_1,y_1) + \frac{\beta^2}{2}\left[\xi(x_1,y_1)\eta_x(x_1,y_1) + \eta(x_1,y_1)\eta_y(x_1,y_1)\right] + \cdots \end{aligned} \tag{1.3-10}$$

or using (1.3-9)

$$
\begin{aligned}
x_2 &= x + \tau\xi(x,y) + \frac{\tau^2}{2}\left[\xi(x,y)\xi_x(x,y) + \eta(x,y)\xi_y(x,y)\right] + \cdots \\
&\quad + \beta\xi\left[x + \tau\xi(x,y) + \cdots,\ y + \tau\eta + \cdots\right] \\
&\quad + \frac{\beta^2}{2}\left[\xi(x,y)\xi_x(x,y) + \eta(x,y)\xi_y(x,y)\right] + \cdots \\
y_2 &= y + \tau\eta(x,y) + \frac{\tau^2}{2}\left[\xi(x,y)\eta_x(x,y) + \eta(x,y)\eta_y(x,y)\right] + \cdots \\
&\quad + \beta\eta\left[x + \tau\xi(x,y) + \cdots,\ y + \tau\eta + \cdots\right] \\
&\quad + \frac{\beta^2}{2}\left[\xi(x,y)\eta_x(x,y) + \eta(x,y)\eta_y(x,y)\right] + \cdots
\end{aligned}
\tag{1.3-11}
$$

Expanding and keeping only the quadratic terms in τ, β we have

$$
\begin{aligned}
x_2 &= x + (\tau+\beta)\xi(x,y) + \frac{(\tau+\beta)^2}{2}\left[\xi(x,y)\xi_x(x,y) + \eta(x,y)\xi_y(x,y)\right] + \cdots \\
y_2 &= y + (\tau+\beta)\eta(x,y) + \frac{(\tau+\beta)^2}{2}\left[\xi(x,y)\eta_x(x,y) + \eta(x,y)\eta_y(x,y)\right] + \cdots
\end{aligned}
$$

This form demonstrates the composition law $\gamma = \tau + \beta$.

In the representation (1.3-9) the identity element corresponds to $\tau = 0$ and the inverse is $-\tau$.

Another way of arriving at the global group starting from the infinitesimal transformations is to consider the formalities of integrations of (1.3-8). The integral curves of the first part of (1.3-8) are of the form

$$\Omega(x_1,y_1) = c = \Omega(x,y) \quad \text{if} \quad x_1 = x, y_1 = y \quad \text{when} \quad \tau = 0. \tag{1.3-12}$$

The constant c thus depends only on the initial point (x,y). Along any such integral curve we have

$$y_1 = y_1(x_1;\ c)$$

so that the parameter τ is found by integration of the second part of (1.3-8), for example,

$$\frac{dx_1}{\xi(x_1, y_1(x_1;\ c))} = d\tau. \qquad (1.3\text{-}13)$$

This, for each c, has an integral of the form

$$F(x_1;\ c) - \tau = \text{const.} \qquad (1.3\text{-}14)$$

In general, $c = c(x,y)$ so that (1.3-14) can be written

$$W(x_1,y_1) - \tau = \text{const.} = W(x,y). \qquad (1.3\text{-}15)$$

The pair (1.3-12, 1.3-15) represent the global form of (1.3-8); $\tau = 0$ is the identity; $(-\tau)$ is the inverse and the composition property follows from

$$\Omega(x_2,y_2) = \Omega(x_1,y_1) = \Omega(x,y)$$

$$W(x_2,y_2) = \tau_1 + W(x_1,y_1) = (\tau_1+\tau) + W(x,y).$$

<u>Summary</u>:

In summary $[\xi(x,y),\ \eta(x,y)]$ define an infinitesimal transformation $\delta x = x_1 - x = \xi\delta\tau$, $\delta y = y_1 - y = \eta\delta\tau$ and this defines a one-parameter group of transformations containing the identity and inverse. The finite form of the group of transformations is found by integration of the differential system

$$\frac{dx_1}{\xi(x_1,y_1)} = \frac{dy_1}{\eta(x_1,y_1)} = d\tau$$

with the initial conditions $x_1 = x$, $y_1 = y$ at $\tau = 0$. The general form of these transformations is

$$\begin{aligned} \Omega(x_1,y_1) &= \Omega(x,y) \\ W(x_1,y_1) &= \tau + W(x,y) \end{aligned} \quad . \tag{1.3-16}$$

If this pair is solved for (x_1,y_1) we obtain formally (ϕ,ψ) and the local series

$$x_1 = \phi(x,y;\ \tau) = x + \tau\xi(x,y) + \frac{\tau^2}{2!}(\xi\xi_x+\eta\xi_y) + \cdots$$

$$y_1 = \psi(x,y;\ \tau) = y + \tau\eta(x,y) + \frac{\tau^2}{2!}(\xi\eta_x+\eta\eta_y) + \cdots$$

The first-order term of this series is identical with the infinitesimal transformation.

Example:

$$\left\{\begin{aligned} \xi(x,y) &= x \\ \eta(x,y) &= y \end{aligned}\right\} \quad \text{or} \quad \left\{\begin{aligned} x_1 - x &= \xi\delta\tau \\ y_1 - y &= \eta\delta\tau \end{aligned}\right\}.$$

The differential system is

$$\frac{dx_1}{x_1} = \frac{dy_1}{y_1} = d\tau$$

which has the integrals satisfying $(x_1 = x,\ y_1 = y,\ \tau = 0)$

$$\log x_1 - \log x = \log y_1 - \log y = \tau, \quad x,\ y,\ x_1,\ y_1 > 0 \ \text{ say.}$$

Parametrically, the global equations are

$$\begin{aligned} x_1 &= \phi(x,y;\ \tau) = xe^{\tau} \\ y_1 &= \psi(x,y;\ \tau) = ye^{\tau} \end{aligned} \quad .$$

Let $e^{\tau} = \alpha$; we have the perspective or similarity transformations

$$x_1 = \alpha x$$
$$y_1 = \alpha y$$

which evidently form a group. The power series corresponding to these transformations is

$$x_1 = x + \tau x + \frac{\tau^2}{2!} x + \cdots = xe^{\tau}$$
$$y_1 = y + \tau x + \frac{\tau^2}{2!} y + \cdots = ye^{\tau}$$

1.4. Proof That a One-Parameter Group Essentially Contains Only One Infinitesimal Transformation and Is Determined by It

In this section an abbreviated notation is used: Denote the pair (x,y) by x and the pair (ϕ,ψ) by ϕ. The results can be generalized to three or more variables.

Let

$$x_1 = \phi(x;\ \alpha) \tag{1.4-1}$$

define a one-parameter group of transformations with $\phi(x;\ 0) = x$ defining the identity transformation. If

$$x_2 = \phi(x_1;\ \beta) \tag{1.4-2}$$

then there is some function $\gamma(\alpha,\beta)$ defining the law of composition such that

$$x_2 = \phi(x;\ \gamma(\alpha,\beta)) \tag{1.4-3}$$

for any values of x,α,β. $\gamma(\alpha,\beta)$ essentially describes how the one-parameter group of transformations is parameterized. $\gamma(\alpha,\beta)$ is an analytic function of α and β in some neighborhood of $(0,0)$. Given α, there is a unique value $\beta = \alpha^{-1}$ (corresponding to the inverse, not necessarily $\frac{1}{\alpha}$) such that $\gamma(\alpha,\alpha^{-1}) = 0$, the identity element. Thus, if

$$x_1 = \phi(x;\ \alpha^{-1}),$$

then

$$x = \phi(x_1;\ \alpha). \tag{1.4-4}$$

We now show that essentially the infinitesimal transformation is unique.

Say α in (1.4-4) is given an infinitesimal increment, $\delta\alpha$. Then

$$x^* = \phi(x_1;\ \alpha+\delta\alpha) = \phi(x_1;\ \alpha) + \delta\alpha \frac{\partial\phi}{\partial\beta}(x_1;\ \beta)\Big|_{\beta=\alpha} + \cdots \tag{1.4-5}$$

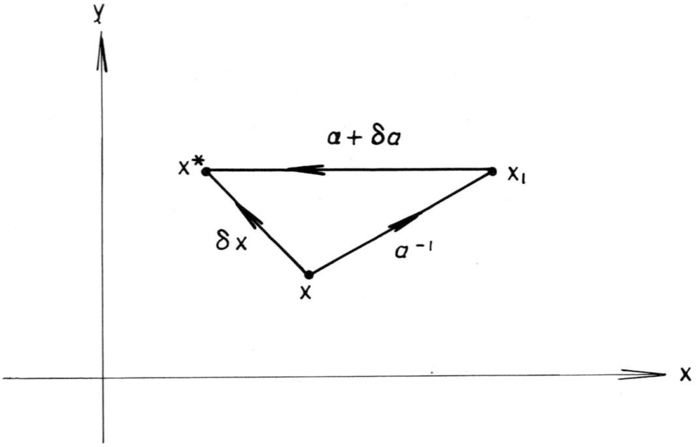

Figure 1.4-1 Geometric interpretation of (1.4-5)

If δx is the infinitesimal change in x, then

$$\delta x = \xi(x;\ \alpha)\,\delta\alpha \tag{1.4-6}$$

where

$$\xi(x;\ \alpha) = \frac{\partial\phi}{\partial\beta}(x_1;\ \beta)\Big|_{\beta=\alpha} \tag{1.4-7}$$

is the infinitesimal of the group of transformations (1.4-1) corresponding to the parameterization $\gamma(\alpha,\beta)$. Note that $\xi(x;\ \alpha) \not\equiv 0$ for all values of α (for otherwise (1.4-1) defines the trivial group).

The group of transformations with law of combination $\gamma(\alpha,\beta)$ can be determined by integrating out the differential equation

$$\frac{dx^*}{\xi(x^*;\ \alpha)} = d\alpha \tag{1.4-8}$$

with initial condition

$$x^*(\alpha = 0) = x. \tag{1.4-9}$$

We show that the infinitesimal is essentially unique by proving that it is of the form

$$\xi(x;\ \alpha) = \Gamma(\alpha)\xi(x) \tag{1.4-10}$$

for some functions $\{\Gamma(\alpha),\ \xi(x)\}$.

<u>Theorem 1.4.1</u>. $\xi(x;\ \alpha) = \Gamma(\alpha)\xi(x)$

where

$$\Gamma(\alpha) = \left.\frac{\partial\gamma(\alpha^{-1},\beta)}{\partial\beta}\right|_{\beta=\alpha} \tag{1.4-11}$$

$$\xi(x) = \xi(x;\ 0) = \left.\frac{\partial\phi(x;\ \alpha)}{\partial\alpha}\right|_{\alpha=0} \tag{1.4-12}$$

<u>Proof</u>. From the definition of $\gamma(\alpha,\beta)$ in the beginning of this section

$$\begin{aligned} x^* &= \phi(x_1;\ \alpha+\delta\alpha) \\ &= \phi(\phi(x;\ \alpha^{-1});\ \alpha+\delta\alpha) \\ &= \phi(x;\ \gamma(\alpha^{-1},\alpha+\delta\alpha)) \end{aligned} \tag{1.4-13}$$

$$\begin{aligned} \gamma(\alpha^{-1},\alpha+\delta\alpha) &= \gamma(\alpha^{-1},\alpha) + \Gamma(\alpha)\delta\alpha + 0((\delta\alpha)^2) \\ &= \Gamma(\alpha)\delta\alpha + 0((\delta\alpha)^2) \end{aligned}$$

since $\gamma(\alpha^{-1},\alpha) = \gamma(\alpha,\alpha^{-1}) = 0$.

Hence

$$\begin{aligned} x^* &= \phi(x;\ \Gamma(\alpha)\delta\alpha + 0((\delta\alpha)^2) \\ &= \phi(x;\ 0) + \delta\alpha\Gamma(\alpha)\left[\frac{\partial\phi}{\partial\alpha}(x;\ \alpha)\right]\Bigg|_{\alpha=0} + 0((\delta\alpha)^2) \\ &= x + \Gamma(\alpha)\xi(x)\delta\alpha + 0((\delta\alpha)^2) \end{aligned}$$

==>

$$\delta x = \Gamma(\alpha)\xi(x)\delta\alpha.$$

Corollary 1.4.1.

$$\xi(x) \not\equiv 0.$$

Problem 1.4.1. Show that $\Gamma(0) = 1$.

(*Hint*: use analyticity property of $\gamma(\alpha,\beta)$).

Letting

$$t = \int_0^\alpha \Gamma(\alpha')d\alpha',$$

we get $\delta x = \xi(x)\delta t$. Hence essentially there is only one infinitesimal.

Examples:

(i) Stretching

$$\begin{aligned} x_1 &= (\alpha+1)x \\ y_1 &= (\alpha+1)y \end{aligned} \tag{1.4-14}$$

$$\gamma(\alpha,\beta) = \alpha\beta + \alpha + \beta$$

$$\alpha^{-1} = -\frac{\alpha}{1+\alpha}$$

$$\frac{\partial\gamma}{\partial\beta}(\alpha,\beta) = \alpha + 1 \quad ==> \quad \Gamma(\alpha) = \frac{\partial\gamma}{\partial\beta}(\alpha^{-1},\beta)\Bigg|_{\beta=\alpha} = \frac{1}{1+\alpha}$$

$$\xi(x) = (x,y)$$

$$\xi(x;\ \alpha) = \left(\frac{x}{1+\alpha},\ \frac{y}{1+\alpha}\right).$$

The corresponding differential equations determining (1.4-14) are

$$\frac{dx^*}{\frac{x^*}{1+\alpha}} = \frac{dy^*}{\frac{y^*}{1+\alpha}} = d\alpha$$

with initial condition $(x^*,y^*)\Big|_{\alpha=0} = (x,y)$.

(ii) Rotation

$$x_1 = \sqrt{1-\alpha^2}\,x - \alpha y$$
$$y_1 = \alpha x + \sqrt{1-\alpha^2}\,y \tag{1.4-15}$$

$$\gamma(\alpha,\beta) = \alpha\sqrt{1-\beta^2} + \beta\sqrt{1-\alpha^2}$$

$$\alpha^{-1} = -\alpha$$

$$\frac{\partial\gamma}{\partial\beta}(\alpha,\beta) = -\frac{\beta\alpha}{\sqrt{1-\beta^2}} + \sqrt{1-\alpha^2}$$

$$\Gamma(\alpha) = \frac{1}{\sqrt{1-\alpha^2}}$$

$$\xi(x) = (-y,x)$$

$$\xi(x;\ \alpha) = \left(-\frac{y}{\sqrt{1-\alpha^2}},\ \frac{x}{\sqrt{1-\alpha^2}}\right).$$

Problem 1.4.2. Integrate out

$$-\sqrt{1-\alpha^2}\ \frac{dx^*}{y^*} = \sqrt{1-\alpha^2}\ \frac{dy^*}{x^*} = d\alpha$$

$$(x^*,y^*)\Big|_{\alpha=0} = (x,y)$$

to obtain (1.4-15).

1.5. Transformations; Symbol of the Infinitesimal Transformation U

In this section the transformation group (which takes the plane into itself) is shown to be independent of its coordinate representation. Further, a useful symbol U for expressing the transformation in terms of its infinitesimals is introduced.

The transformation of the plane into itself can be expressed in any coordinates. The group property is preserved independent of the choice of coordinates. All one-parameter groups in the plane can be brought to the same form by a suitable choice of coordinates. To show these results in more detail consider $(\tilde{x},\tilde{y})$

$$\begin{aligned} \tilde{x} &= F(x,y) \\ \tilde{y} &= G(x,y) \end{aligned} \tag{1.5-1}$$

not as a transformation but rather as new coordinates in the plane.

Example: polar coordinates $\tilde{x} = \sqrt{x^2+y^2}$, $\tilde{y} = \tan^{-1}y/x$.

If the general transformation group is

$$\begin{aligned} x_1 &= \phi(x,y;\ \alpha) \\ y_1 &= \psi(x,y;\ \alpha) \end{aligned} \tag{1.5-2}$$

the new point (x_1,y_1) can also be represented as

$$\begin{aligned} \tilde{x}_1 &= F(x_1,y_1) \\ \tilde{y}_1 &= G(x_1,y_1) \end{aligned}$$

Replacing (x_1,y_1) by (1.5-2) and (x,y) by (1.5-1) inverted, we have

$$\begin{aligned} \tilde{x}_1 &= \Phi(\tilde{x},\tilde{y};\ \alpha) \\ \tilde{y}_1 &= \Psi(\tilde{x},\tilde{y};\ \alpha) \end{aligned} \tag{1.5-3}$$

which is a new representation of the group.

Example: A one-parameter family of transformations is

$$x_1 = x + \alpha$$

$$y_1 = \frac{xy}{x+\alpha}$$

The group composition property is verified by

$$x_2 = x_1 + \beta = x + (\alpha+\beta)$$

$$y_2 = \frac{x_1y_1}{x_1+\beta} = \frac{(x+\alpha)\frac{xy}{x+\alpha}}{x+(\alpha+\beta)}$$

$$= \frac{xy}{x+(\alpha+\beta)}$$

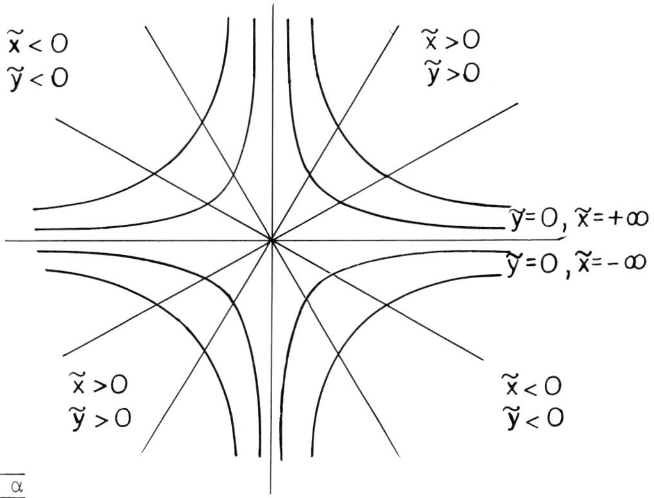

Figure 1.5-1

$\alpha = 0$ is the identity, $-\alpha$ is the inverse of α. Choose as new coordinates rays and hyperbolas

$$\left\langle \begin{matrix} \tilde{x} = \frac{x}{y} \\ \tilde{y} = xy \end{matrix} \right\rangle \qquad \begin{matrix} x = \pm\sqrt{\tilde{x}\tilde{y}} \\ y = \pm\sqrt{\tilde{y}/\tilde{x}} \end{matrix}$$

- the four choices of sign cover the plane.

Thus, in new coordinates the basic transformation is

$$\tilde{x}_1 = \frac{x_1}{y_1} = \frac{(\pm\sqrt{\tilde{x}\tilde{y}} + \alpha)^2}{\tilde{y}}$$

$$\tilde{y}_1 = x_1y_1 = xy = \tilde{y}\ .$$

The group property in new coordinates is

$$\tilde{x}_2 = \frac{(\pm\sqrt{\tilde{x}_1\tilde{y}_1} + \beta)^2}{\tilde{y}_1} = \frac{\left[\pm\sqrt{(\pm\sqrt{\tilde{x}\tilde{y}} + \alpha)^2} + \beta\right]^2}{\tilde{y}} = \frac{(\pm\sqrt{\tilde{x}\tilde{y}} + (\alpha+\beta))^2}{\tilde{y}} .$$

Theoretically, new coordinates can be chosen so that the group has the form of a translation. Referring to (1.3-16) we can choose canonical coordinates

$$\begin{aligned} r &= \Omega(x,y) \\ s &= W(x,y) \end{aligned} \tag{1.5-4}$$

so that the group is

$$\begin{aligned} r_1 &= r \\ s_1 &= s + \tau \end{aligned} . \tag{1.5-5}$$

For the above example $r = xy$, $s = x$.

U-Symbol of the Infinitesimal Transformation

The symbol U is introduced as the symbol for a directional derivative in the plane (space, etc.). The use of this symbol facilitates calculation and separates the role of variables $(x,y\ldots)$ and parameters (α,β).

Consider the one-parameter group of transformations

$$\begin{aligned} x_1 &= \phi(x,y;\ \tau) \\ y_1 &= \psi(x,y;\ \tau) \end{aligned} \tag{1.5-6}$$

and the corresponding infinitesimal transformation

$$\begin{aligned} x^* &= x_1 + \xi(x_1,y_1)\delta\tau \\ y^* &= y_1 + \eta(x_1,y_1)\delta\tau \end{aligned} . \tag{1.5-7}$$

Now consider how a function $f(x^*,y^*)$ defined over the plane varies along the path curve of a given (arbitrary) initial point (x,y).

$$\delta f_1 = f(x^*,y^*) - f(x_1,y_1) = \frac{\partial f}{\partial x_1}(x_1,y_1)\delta x_1 + \frac{\partial f}{\partial y_1}(x_1,y_1)\delta y_1$$

$$= \left[\xi(x_1,y_1)\frac{\partial f}{\partial x_1}(x_1,y_1) + \eta(x_1,y_1)\frac{\partial f}{\partial y_1}(x_1,y_1)\right]\delta\tau$$

in particular as $\tau \to 0$; we approach the initial point (x,y) so

$$\lim_{\tau \to 0}\frac{\delta f}{\delta\tau} = \frac{df}{d\tau}\ (\text{along the path}) = \xi(x,y)\frac{\partial f}{\partial x} + \eta(x,y)\frac{\partial f}{\partial y} \qquad (1.5\text{-}8)$$

$$\tau = 0$$

(1.5-8) is the usual directional derivative along the path

$$\xi\cdot\nabla \equiv \xi\frac{\partial}{\partial x} + \eta\frac{\partial}{\partial y}\ .$$

This operator is labeled U:

$$Uf \equiv \xi(x,y)\frac{\partial f}{\partial x} + \eta(x,y)\frac{\partial f}{\partial y}\ . \qquad (1.5\text{-}9)$$

Example: Rotation: infinitesimally

$$\delta x = x_1 - x = -y\delta\tau$$

$$\delta y = y_1 - y = x\delta\tau$$

$$Uf \equiv -y\frac{\partial f}{\partial x} + x\frac{\partial f}{\partial y}$$

$$\xi = -y, \qquad \eta = x.$$

Note that

$$\xi = Ux, \qquad \eta = Uy$$

so that in general

$$Uf \equiv \xi\frac{\partial f}{\partial x} + \eta\frac{\partial f}{\partial y} \equiv (Ux)\frac{\partial f}{\partial x} + (Uy)\frac{\partial f}{\partial y} \qquad (1.5\text{-}10)$$

$$\text{where}\quad (\xi,\eta) \to U.$$

The directional derivative expressed by (1.5-10) can easily be transformed to new coordinates. Let

$$\begin{aligned} \tilde{x} &= F(x,y) \\ \tilde{y} &= G(x,y) \end{aligned} \tag{1.5-11}$$

be new coordinates. Then the new coordinates of a point on a path curve are

$$\tilde{x}_1 = F(x_1,y_1) = F(x+\xi\delta\tau, y+\eta\delta\tau) = F(x,y) + (\xi F_x+\eta F_y)\delta\tau + \cdots$$

$$\tilde{y}_1 = G(x_1,y_1) = G(x+\xi\delta\tau, y+\eta\delta\tau) = G(x,y) + (\xi G_x+\eta G_y)\delta\tau + \cdots$$

or

$$\begin{aligned} \delta\tilde{x} &= \tilde{x}_1 - \tilde{x} = (\xi F_x+\eta F_y)\delta\tau \\ \delta\tilde{y} &= \tilde{y}_1 - \tilde{y} = (\xi G_x+\eta G_y)\delta\tau \end{aligned} \quad . \tag{1.5-12}$$

For any function $h(\tilde{x},\tilde{y})$ the relation $\delta h = (\tilde{U}h)\delta\tau$ infinitesimally defines $\tilde{U}$. That is,

$$\tilde{U}h = \frac{\partial h}{\partial \tilde{x}} \frac{\delta\tilde{x}}{} + \frac{\partial h}{\partial \tilde{y}} \frac{\delta\tilde{y}}{\delta\tau} = \left(\xi \frac{\partial F}{\partial x} + \eta \frac{\partial F}{\partial y}\right) \frac{\partial h}{\partial \tilde{x}} + \left(\xi \frac{\partial G}{\partial x} + \eta \frac{\partial G}{\partial x}\right) \frac{\partial h}{\partial \tilde{y}}$$

or the infinitesimal transformation is expressed in new coordinates $\tilde{x}(x,y)$, $\tilde{y}(x,y)$ by

$$\tilde{U}h = (U\tilde{x}) \frac{\partial h}{\partial \tilde{x}} + (U\tilde{y}) \frac{\partial h}{\partial \tilde{y}} \ . \tag{1.5-13}$$

Sometimes the symbol f is used for a function defined over the plane (a certain value attached to each point of the plane) independent of the coordinates, in which case

$$Uf \equiv \xi \frac{\partial f}{\partial x} + \eta \frac{\partial f}{\partial y} = (U\tilde{x}) \frac{\partial f}{\partial \tilde{x}} + (U\tilde{y}) \frac{\partial f}{\partial \tilde{y}} \tag{1.5-14}$$

where $\tilde{x} = \tilde{x}(x,y)$, $\tilde{y} = \tilde{y}(x,y)$

$$\tilde{\xi} = U\tilde{x} = \xi\tilde{x}_x + \eta\tilde{x}_y, \quad \tilde{\eta} = U\tilde{y} = \xi\tilde{y}_x + \eta\tilde{y}_y \tag{1.5-15}$$

defines the same group in new coordinates $(\tilde{x},\tilde{y})$: $\tilde{\xi}(\tilde{x},\tilde{y})$, $\tilde{\eta}(\tilde{x},\tilde{y}) \to \tilde{U}$.

Alternatively, the same result is expressed by the formal rule

$$\begin{aligned} Uf &= \xi \frac{\partial f}{\partial x} + \eta \frac{\partial f}{\partial y} = \xi\left(\frac{\partial f}{\partial \tilde{x}}\frac{\partial \tilde{x}}{\partial x} + \frac{\partial f}{\partial \tilde{y}}\frac{\partial \tilde{y}}{\partial x}\right) + \eta\left(\frac{\partial f}{\partial \tilde{x}}\frac{\partial \tilde{x}}{\partial y} + \frac{\partial f}{\partial \tilde{y}}\frac{\partial \tilde{y}}{\partial y}\right) \\ &= \left(\xi\tilde{x}_x + \eta\tilde{x}_y\right)\frac{\partial f}{\partial \tilde{x}} + \left(\xi\tilde{y}_x + \eta\tilde{y}_y\right)\frac{\partial f}{\partial \tilde{y}} \equiv \tilde{U}f \end{aligned} \tag{1.5-16}$$

(this corresponds to the usual physicist's notation , f and the operator here should be understood that way).

<u>Example</u>: polar coordinates $x = r\cos\theta$, $y = r\sin\theta$

$$\tilde{x} = \sqrt{x^2+y^2}$$

$$\tilde{y} = \theta = \tan^{-1}\frac{y}{x}\ .$$

Let $f \equiv x\sqrt{x^2+y^2} = r^2\cos\theta$ in polar coordinates. Let the transformation be rotation $Uf = -y\,\partial f/\partial x + x\,\partial f/\partial y$. Thus, in (x,y)

$$Uf = -y\left[\sqrt{x^2+y^2} + \frac{x^2}{\sqrt{x^2+y^2}}\right] + x\left[\frac{xy}{\sqrt{x^2+y^2}}\right] = -y\sqrt{x^2+y^2}\ .$$

Computing U in $(\tilde{x},\tilde{y})$ we have

$$\tilde{U}f = (U\tilde{x})\frac{\partial f}{\partial \tilde{x}} + (U\tilde{y})\frac{\partial f}{\partial \tilde{y}}\ ; \quad U\tilde{x} = -y\frac{\partial}{\partial x}\sqrt{x^2+y^2} + x\frac{\partial}{\partial y}\left(\sqrt{x^2+y^2}\right) = 0$$

$$U\tilde{y} = -y\left[\frac{-y/x^2}{1+(y/x)^2}\right] + x\left[\frac{1/x}{1+(y/x)^2}\right] = 1.$$

Thus, $\tilde{U}f = \dfrac{\partial f}{\partial \tilde{y}} = \dfrac{\partial f}{\partial \theta} = -r^2\sin\theta = -y\sqrt{x^2+y^2}$ as before.

Note that, in general, the infinitesimal transformation can be used to provide conditions for canonical coordinates of a given group. In canonical coordinates $r = \tilde{x}$, $s = \tilde{y}$, the group is a translation, as in the example above, so that

$$\tilde{U}f \equiv \frac{\partial f}{\partial s} \quad \text{or} \quad \begin{aligned} r_1 &= r \\ s_1 &= s + \alpha \end{aligned} . \tag{1.5-17}$$

Thus,

$$\begin{aligned} \tilde{\xi} &= 0 = \xi \frac{\partial r}{\partial x} + \eta \frac{\partial r}{\partial y} \\ \tilde{\eta} &= 1 = \xi \frac{\partial s}{\partial x} + \eta \frac{\partial s}{\partial y} \end{aligned} . \tag{1.5-18}$$

An explicit determination of the canonical coordinates depends on an integration of the system (1.5-18). The characteristic system of (1.5-18) is

$$\frac{dx}{\xi(x,y)} = \frac{dy}{\eta(x,y)} = \frac{dr}{0}; \; r(x,y) = c = \text{const. on curves whose slope is } \eta/\xi. \tag{1.5-19}$$

$$\frac{dx}{\xi(x,y)} = \frac{dy}{\eta(x,y)} = \frac{ds}{1}; \; s = \int \frac{dx}{\xi(x;\ y(x;\ r))} \tag{1.5-20}$$

$s = s(x,y)$ is found by quadrature.

<u>Example: Stretching Group</u> $\quad \begin{aligned} x_1 &= \alpha x \\ y_1 &= \alpha^k y \end{aligned}$

Infinitesimally:

$$\alpha = 1 + \delta\tau$$

$$\alpha^k = 1 + k\delta\tau$$

$$Uf \equiv x \frac{\partial f}{\partial x} + ky \frac{\partial f}{\partial y} \quad \text{or} \quad \xi = x, \; \eta = ky.$$

Canonical coordinates:

$$0 = x\frac{\partial r}{\partial x} + ky\frac{\partial r}{\partial y}$$

$$1 = x\frac{\partial s}{\partial x} + ky\frac{\partial s}{\partial y}.$$

Characteristic equations:

$$\frac{dx}{x} = \frac{dy}{ky} = \frac{dr}{0} = \frac{ds}{1}.$$

Integral curves:

$$k \log x - \log y = \text{const.} \quad x, y > 0.$$

So $r = F(y/x^k)$, where F is an arbitrary function, is a canonical coordinate. For simplicity let

$$r = y/x^k \quad x, y > 0.$$

To find the other canonical coordinate $s(x,y)$ only a particular solution is needed, say

$$s = \log x, \quad x > 0;$$

in these coordinates

$$r_1 = \frac{y_1}{x_1^k} \doteq \frac{\alpha^k y}{\alpha^k x^k} = \frac{y}{x^k} = r$$

$$s_1 = \log x_1 = \log x + \log \alpha$$
$$= s + \log \alpha$$

$$Uf \equiv \frac{\partial f}{\partial s}.$$

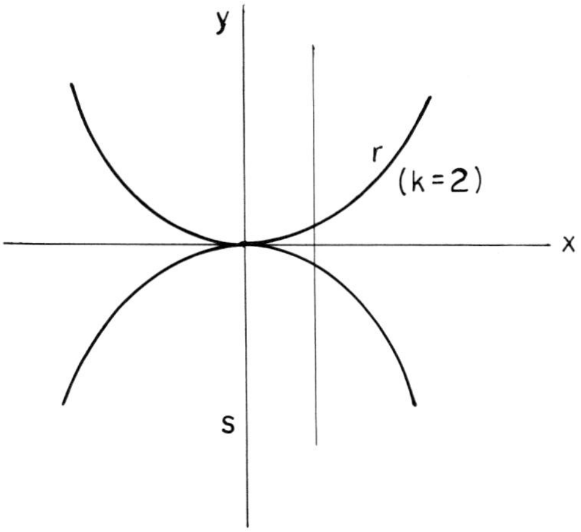

Figure 1.5-2

Series for the Group

Let x_1, y_1 be on the path curve of (x,y), that is these points are given by (1.5-6). Any function defined over the plane can be considered along this path curve

$$g(x_1, y_1) \text{ is a function of } (x,y,\tau).$$

The rate of change of g along the path curve can be expressed near $\tau = 0$ (assumed the identity) by

$$\left(\frac{dg}{d\tau}\right)_{\tau=0}$$

and in fact, for some finite interval around $\tau = 0$ we can write

$$g\left[x_1(\tau),\ y_1(\tau)\right] = g\left[x_1(0),\ y_1(0)\right] + \tau\left(\frac{dg}{d\tau}\right)_{\tau=0} + \frac{\tau^2}{2!}\left(\frac{d^2g}{d\tau^2}\right)_{\tau=0} + \cdots \tag{1.5-21}$$

but

$$\frac{dg}{d\tau} \equiv \frac{\partial g}{\partial x}\frac{dx_1}{d\tau} + \frac{\partial g}{\partial y}\frac{dy_1}{d\tau} = \xi_1\left(\frac{\partial g}{\partial x}\right)_1 + \eta_1\left(\frac{\partial g}{\partial y}\right)_1 .$$

At $\tau = 0$

$$\left(\frac{dg}{d\tau}\right)_{\tau=0} = \xi(x,y)\frac{\partial g}{\partial x} + \eta(x,y)\frac{\partial g}{\partial y} = Ug. \tag{1.5-22}$$

The process continues in an obvious manner to higher derivatives, i.e.,

$$\left(\frac{d^2g}{dt^2}\right)_{\tau=0} = U^2g, \qquad \left(\frac{d^3g}{dt^3}\right)_{\tau=0} = U^3g \quad \text{etc.}$$

Thus, for any function, the changes along a path are expressed by a series; (1.5-21) reads

$$\begin{aligned} g(x_1,y_1) &= g(x,y) + \tau\, Ug + \frac{\tau^2}{2!}U^2g + \cdots \\ &= e^{\tau U}g(x,y) \qquad \text{formally.} \end{aligned} \tag{1.5-23}$$

(1.5-23) can be applied to (x,y) to obtain the series representation of the global transformation (1.5-6)

$$x_1 = x + \tau\, Ux + \frac{\tau^2}{2!} U^2x + \cdots, \quad Ux = \xi$$
$$y_1 = y + \tau\, Uy + \frac{\tau^2}{2!} U^2y + \cdots, \quad Uy = \eta \tag{1.5-24}$$

so that

$$x_1 = x + \tau\xi + \frac{\tau^2}{2!} (\xi\xi_x+\eta\xi_y) + \cdots$$
$$y_1 = y + \tau\eta + \frac{\tau^2}{2!} (\xi\eta_x+\eta\eta_y) + \cdots \tag{1.5-25}$$

The series representation (1.5-24) splits off the dependence on τ from (x,y) calculations. Repeated application of the directional derivative takes us along the path.

<u>Example</u>: (i) infinitesimal rotation $\xi = -y$, $\eta = x$

$$Uf = -y \frac{\partial f}{\partial x} + x \frac{\partial f}{\partial y}$$
$$Ux = -y, \quad Uy = x$$
$$U^2x = -Uy = -x, \quad U^2y = -y$$
$$U^3x = -Ux = y, \quad U^3y = -x$$
$$U^4x = Uy = x, \quad U^4y = y$$

$$x_1 = x + \tau(-y) + \frac{\tau^2}{2!}(-x) + \frac{\tau^3}{3!}(y) + \frac{\tau^4}{4!}(x) + \cdots = x \cos\tau - y \sin\tau$$
$$y_1 = y + \tau(x) + \frac{\tau^2}{2!}(-y) + \frac{\tau^3}{3!}(-x) + \frac{\tau^4}{4!}(y) + \cdots = y \cos\tau + x \sin\tau$$

<u>Exercise</u>: Find the global groups corresponding to:

(ii) $Uf \equiv x \frac{\partial f}{\partial x} + y \frac{\partial f}{\partial y}$

(iii) $Uf \equiv x \frac{\partial f}{\partial x}$

(iv) $Uf = x^2 \frac{\partial f}{\partial x} + xy \frac{\partial f}{\partial y}$

(v) $Uf = e^x \frac{\partial f}{\partial x}$.

1.6. Invariant Functions and Curves

In this section, a definition of invariance is formulated which is useful for application to ordinary and later partial differential equations. A one-parameter group of transformations is defined by U or the global equations and contains (∞^1) path curves. As the transformation is applied a representative point (x_1,y_1) moves along a path curve so that the path curve goes into itself - is invariant. This concept is to be expressed analytically.

Curves in the plane can be expressed by $\Omega(x,y) = \text{const.}$ so that we consider first a definition of invariance for a function $\Omega(x,y)$ defined over the plane:

$$\Omega(x,y) \text{ is invariant iff } \Omega(x_1,y_1) = \Omega(x,y) \text{ for all values of } \tau. \tag{1.6-1}$$

For example, under rotation around the origin $x^2 + y^2 = x_1^2 + y_1^2$, so that $\Omega \equiv x^2 + y^2$ is an invariant function. A local differential condition for invariance is found from (1.5-23).

$$\Omega(x_1,y_1) = \Omega(x,y) + \tau U\Omega(x,y) + \frac{\tau^2}{2!} U^2\Omega(x,y) + \cdots . \tag{1.6-2}$$

Evidently, a necessary and sufficient condition for invariance is

$$U\Omega = 0 \quad \text{for all} \quad (x,y),$$

that is $\Omega(x,y)$ is a solution of the partial differential equation

$$\xi(x,y) \frac{\partial \Omega}{\partial x} + \eta(x,y) \frac{\partial \Omega}{\partial y} = 0. \tag{1.6-3}$$

Note that if one invariant Ω is found any function $F(\Omega)$ is also an invariant. Note also that $\Omega(x,y) = \text{const.}$ defines a path curve when $\Omega(x,y)$ is an invariant function since

$$d\Omega = 0 = \Omega_x dx + \Omega_y dy \quad \text{on} \quad \Omega = \text{const.}$$

$$\left(\frac{dy}{dx}\right)_{\Omega=\text{const.}} = -\frac{\Omega_x}{\Omega_y} = \frac{\eta(x,y)}{\xi(x,y)} . \tag{1.6-4}$$

Note also that the (∞^1) path curves are determined explicitly in the form

$$W(x_1,y_1;\ x,y) = 0 \tag{1.6-5}$$

by elimination of the parameter τ from the global transformation (1.5-6). Since (x_1,y_1) lies on the path only one of these is free so that (1.6-5) is of the form

$$\Omega(x,y) = \text{const.} = c. \tag{1.6-6}$$

In a similar way <u>invariant</u> curves in the plane are defined as curves which are unchanged as (all) members of the one-parameter group of transformations are applied to the plane. This can happen in either of two ways

(i) each point of the curve does not move

(ii) each point of the curve moves along the curve (path curves).

Since the transformations are given by (1.5-24) the condition that points do not move under U

$$Uf \equiv \xi(x,y)\frac{\partial f}{\partial x} + \eta(x,y)\frac{\partial f}{\partial y}$$

is

$$\text{(i)} \quad \xi(x,y) = 0,\ \eta(x,y) = 0 \quad \text{for curves formed of invariant points.} \tag{1.6-7}$$

Otherwise, a <u>curve is invariant</u> if

when $\omega(x,y) = 0$ describes the curve

$\omega(x_1,y_1) = 0,$ or the curve goes to itself.

It follows again from (1.5-23) that

$$\boxed{U\omega = 0 \quad \text{when} \quad \omega = 0}\ , \quad \underline{\text{invariance condition}}. \tag{1.6-8}$$

This is the basic condition of invariance for a curve.

Note, however, that (1.6-8) should not be satisfied trivially, that is $\omega(x,y)$ should not be written so that both $(\omega_x,\omega_y) = 0$. Such curves have the tangent direction of the transformation as in (1.6-4) and are path curves. Also, note that only a knowledge of η/ξ is required to find the path curves.

<u>In summary</u>: Two types of curves remain invariant under the one-parameter group U

(1) path curves: defined by invariant functions $\Omega(x,y) = \text{const.}$ or $\omega(x,y) = 0$;

(2) curves composed of invariant points $\xi(x,y) = 0,\ \eta(x,y) = 0$. The necessary and sufficient condition is

$$U\omega = 0 \quad \text{when} \quad \omega = 0 \quad [\text{not both } \omega_x,\omega_y = 0].$$

<u>Example</u>:

$$Uf \equiv y\,\frac{\partial f}{\partial x} \qquad \xi = y,\ \eta = 0.$$

The global group is given by

$$Ux = y,\ U^2x = 0,\ Uy = 0 \quad \left\{ \begin{aligned} x_1 &= x + y\tau \\ y_1 &= y \end{aligned} \right.$$

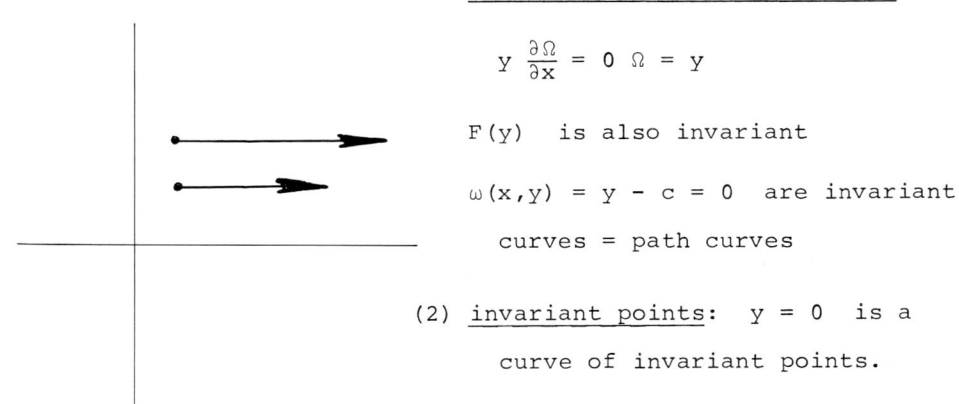

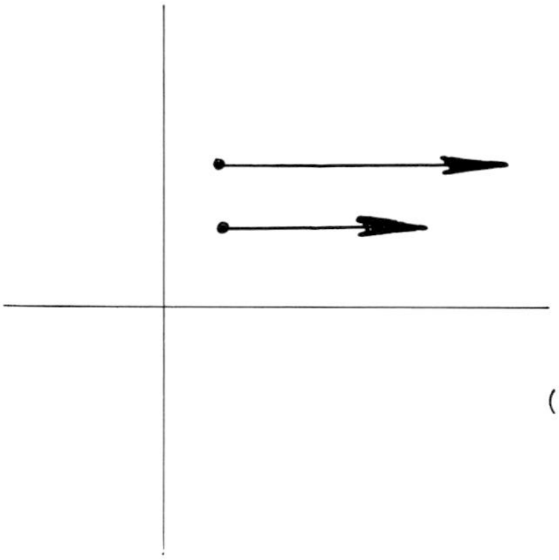

(1) invariant function and curves:

$$y \frac{\partial \Omega}{\partial x} = 0 \quad \Omega = y$$

$F(y)$ is also invariant

$\omega(x,y) = y - c = 0$ are invariant curves = path curves

(2) invariant points: $y = 0$ is a curve of invariant points.

Figure 1.6-1

Note that the fact that a curve is invariant or "admits" a group of transformations does not depend on the coordinate system used to express the curve or the transformations.

Problem 1.6.1. Find invariant functions and curves for

(i) $Uf = x^2 \frac{\partial f}{\partial x} + y \frac{\partial f}{\partial y}$

(ii) $Uf = x \frac{\partial f}{\partial x} - y \frac{\partial f}{\partial y}$

(iii) $Uf \equiv x^2 \frac{\partial f}{\partial x} + y^2 \frac{\partial f}{\partial y}$

Problem 1.6.2. Show $x^2 + y^2 = 1$ is invariant under

$$Uf \equiv \left(x - \frac{1}{2x}\right) \frac{\partial f}{\partial x} + \left(y - \frac{1}{2y}\right) \frac{\partial f}{\partial y} .$$

1.7. Important Classes of Transformations

In this section are summarized the properties of some of the more commonly occurring and useful transformations.

1. Projective Transformations in the Plane

Projective transformations of the plane into itself are characterized by the mapping of all straight lines into straight lines. They are of the form

$$x_1 = \frac{\alpha x + \beta y + \gamma}{\varepsilon x + \zeta y + \theta}$$
$$y_1 = \frac{\kappa x + \lambda y + \mu}{\varepsilon x + \zeta y + \theta} \tag{1.7-1}$$

with 9 parameters $\alpha, \beta, \gamma, \varepsilon, \zeta, \theta, \kappa, \lambda, \mu$ (8 independent).

Exercise: Verify that straight lines go into straight lines.

To study this infinitesimally note that the identity element has $\alpha = 1$, $\lambda = 1$, $\theta = 1$ and all other parameters equal to zero. Thus let

$$\begin{array}{lll} \alpha \to 1 + \delta\alpha & \beta \to \delta\beta & \gamma \to \delta\gamma \\ \lambda \to 1 + \delta\lambda & \varepsilon \to \delta\varepsilon & \zeta \to \delta\zeta \\ \theta \to 1 + \delta\theta & \kappa \to \delta\kappa & \mu \to \delta\mu \end{array}$$

Then

$$x_1 = \frac{x(1+\delta\alpha) + y\delta\beta + \delta\gamma}{x\delta\varepsilon + y\delta\zeta + 1 + \delta\theta}$$
$$y_1 = \frac{x\delta\kappa + y(1+\delta\lambda) + \delta\mu}{x\delta\varepsilon + y\delta\zeta + 1 + \delta\theta}$$

or, now if $\delta\kappa \to \kappa\delta\tau$ etc. then

$$\delta x = x_1 - x = [\gamma + (\alpha-\theta)x + \beta y - \zeta xy - \varepsilon x^2]\,\delta\tau$$
$$\delta y = y_1 - y = [\mu + \kappa x + (\lambda-\theta)y - \zeta y^2 - \varepsilon xy]\,\delta\tau \quad . \tag{1.7-2}$$

There are basically eight independent one-parameter groups, which of course can be combined infinitesimally linearly. For these eight, the infinitesimal generators Uf are

$$\frac{\partial f}{\partial x}, \frac{\partial f}{\partial y}, x\frac{\partial f}{\partial x}, y\frac{\partial f}{\partial y}, x\frac{\partial f}{\partial y}, y\frac{\partial f}{\partial x},$$
$$x^2\frac{\partial f}{\partial x} + xy\frac{\partial f}{\partial y}, xy\frac{\partial f}{\partial x} + y^2\frac{\partial f}{\partial y}. \qquad (1.7\text{-}3)$$

Invariant points occur for those (x,y) in (1.7-2) which make $\delta x = \delta y = 0$. In general, this gives only a cubic equation for x (show as an exercise) so that three points in the plane and only three straight lines remain invariant, in general. That is, each projective transformation leaves a triangle invariant. As an example, we can choose coordinates so that the triangle is formed of the (x,y) axes and a line at infinity. Thus

$$\xi = \beta x, \quad \eta = \theta y. \qquad (1.7\text{-}4)$$

The path curves are found by integrating

$$\frac{dx_1}{\beta x_1} = \frac{dy_1}{\theta y_1} = d\tau \quad \text{with} \quad x_1 = x, \quad y_1 = y \quad \text{at} \quad \tau = 0, \qquad (1.7\text{-}5)$$

or

$$\left.\begin{aligned} x_1 &= xe^{\beta\tau} \\ y_1 &= ye^{\theta\tau} \end{aligned}\right\} \text{parametrically} \qquad (1.7\text{-}6)$$

or

$$\left(\frac{x_1}{x}\right)^{\theta} = \left(\frac{y_1}{y}\right)^{\beta} \qquad (1.7\text{-}7)$$

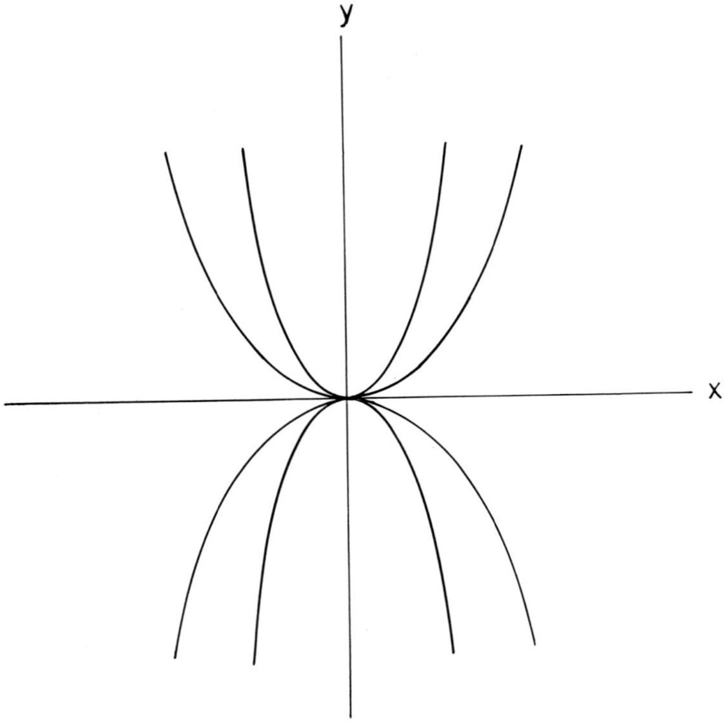

Projective Path Curves
$\theta = 2, \ \beta = 1$

Figure 1.7-1

2. Conformal Transformation (Angle Preserving)

By function theory we know

$$z_1 = x_1 + iy_1 = F(z;\ \tau), \quad z = x + iy \tag{1.7-8}$$

is the global equation of a group in the plane if $F(z;\ \tau)$ is an analytic function of z.

Infinitesimally

$$z_1 - z = \delta z = F_\tau(z;\ 0)\,\delta\tau = (\xi+i\eta)\,\delta\tau \tag{1.7-9}$$

(ξ,η) are the real and imaginary parts of an analytic function of z so that the Cauchy-Riemann equations give us

$$\xi_x = \eta_y, \quad \xi_y = -\eta_x \ . \tag{1.7-10}$$

All transformations for which (1.7-10) hold are conformal.

Special projective transformations are conformal, i.e., when

$$Uf \equiv (\alpha+\beta x+\gamma y)\frac{\partial f}{\partial x} + (\varepsilon-\gamma x+\beta y)\frac{\partial f}{\partial y} \ . \tag{1.7-11}$$

These include:

$$\text{translations: } \frac{\partial f}{\partial x}, \frac{\partial f}{\partial y} \qquad \text{stretching: } x\frac{\partial f}{\partial x} + y\frac{\partial f}{\partial y}$$

$$\text{rotation: } -y\frac{\partial f}{\partial x} + x\frac{\partial f}{\partial y} \ .$$

3. Area Preserving Transformations

Let the one-parameter group be represented temporarily by the vector

$$\underline{x_1} = \underline{\phi}(x;\ \tau) = \underline{x} + \underline{\xi}(x)\,\delta\tau \tag{1.7-12}$$

Let some infinitesimal area at $P(\underline{x})$ be defined by the (arbitrary) infinitesimal vectors $(\underline{\delta v}, \underline{\delta w})$ and thus be proportional to $|\underline{\delta v} \times \underline{\delta w}|$

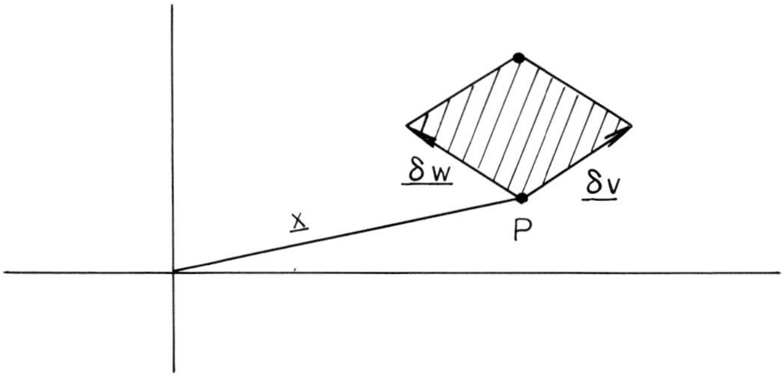

Figure 1.7-2

Then

$$\underline{x_1} + \underline{\delta v_1} = \underline{x} + \underline{\delta v} + \underline{\xi}(\underline{x}+\underline{\delta v})\,\delta\tau = \underline{x} + \underline{\delta v} + \underline{\xi}(\underline{x})\,\delta\tau + \underline{\delta v}\cdot\underline{\nabla\xi}\ \delta\tau$$

$$\underline{\delta v_1} = \underline{\delta v} + \underline{\delta v}\cdot\underline{\nabla\xi}\ \delta\tau$$

$$\underline{x_1} + \underline{\delta w_1} = \underline{x} + \underline{\delta w} + \underline{\xi}(\underline{x})\,\delta\tau + \underline{\delta w}\cdot\underline{\nabla\xi}\ \delta\tau$$

$$\underline{\delta w_1} = \underline{\delta w} + \underline{\delta w}\cdot\underline{\nabla\xi}\ \delta\tau .$$

Thus

$$\underline{\delta v_1} \times \underline{\delta w_1} = (\underline{\delta v} \times \underline{\delta w})\{1 + \text{div}\ \underline{\xi}\ \delta\tau\}. \tag{1.7-12}$$

Area preservation occurs when

$$\text{div}\ \underline{\xi} = \frac{\partial\xi}{\partial x} + \frac{\partial\eta}{\partial y} = 0.$$

Thus, if (ξ,η) are interpreted as the velocity components of any steady incompressible fluid flow in the plane a stream function $\Psi(x,y)$ exists so that

$$\xi = \Psi_y, \qquad \eta = -\Psi_x$$

and the path curves are $\Psi(x,y) = \text{const.}$ (Note: the path curves of an arbitrary transformation can be interpreted as the streamlines of a compressible steady flow.)

1.8. Applications to Differential Equations; Invariant Families of Curves

Differential equations define families of curves (first-order, ∞^1). Thus, we study families of curves which remain invariant or "admit" a given group, in order to prepare a general theory.

The idea is that a family of curves is invariant if for a curve in the family

$$\omega(x,y) = \text{const.} = c \tag{1.8-1}$$

then

$$\omega(x_1,y_1) = \text{const.} = c_1 \text{ (i.e, is a member of the same family)}. \quad (1.8\text{-}2)$$

Example: $\omega(x,y) = y - \kappa x = b$ is a family of straight lines (fix κ, b varies). Under the transformation group (parameter μ)

$$x_1 = x + \mu$$

$$y_1 = y + \mu\lambda$$

we have

$$\omega(x_1,y_1) = y_1 - \kappa x_1 = y + \mu\lambda - \kappa x - \mu\kappa = b + \mu(\lambda-\kappa).$$

Each member of the family is mapped into another member of the family if $\lambda \neq \kappa$; if $\lambda = \kappa$ each line is mapped onto itself.

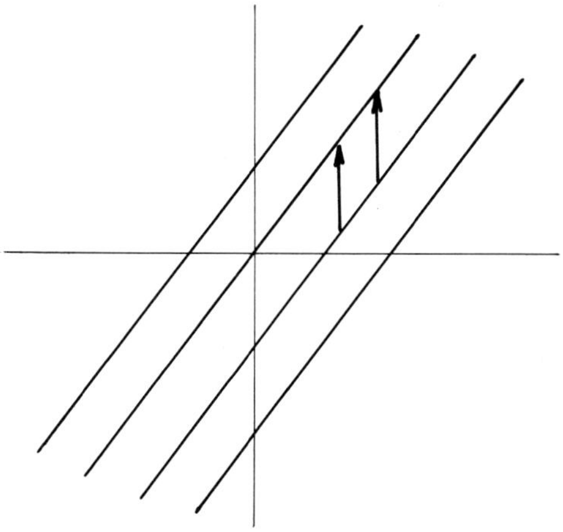

Figure 1.8-1

Now, in general, the invariance criterion can be expressed analytically if we first note that if

$$\omega(x,y) = c$$

$$\sigma(x,y) = k$$

are two representations of the same family then each curve of one family (c) is identical with a curve of the second family (k). A relation exists

$$k = f(c). \tag{1.8-3}$$

It is useful to express the parameters (c,k) by coordinates. Then (1.8-3) becomes

$$\sigma(x,y) = f(\omega(x,y)). \tag{1.8-4}$$

Thus, if (1.8-1) represents a family of (∞^1) curves and if this family admits the transformation

$$\left\{\begin{array}{l} x_1 = \phi(x,y) \\ y_1 = \psi(x,y) \end{array}\right\} \quad \text{or inversely} \quad \left\{\begin{array}{l} x = \Phi(x_1,y_1) \\ y = \Psi(x_1,y_1) \end{array}\right\} \tag{1.8-5}$$

then the condition for invariance is that

$$\omega(\Phi(x_1,y_1),\ \Psi(x_1,y_1)) = \text{const.}$$

which is another representation of

$$\omega(x_1,y_1) = \text{const.}$$

Thus, according to (1.8-4), for invariance, a function must exist such that

$$\omega\left[\Phi(x_1,y_1),\ \Psi(x_1,y_1)\right] = f\left[\omega(x_1,y_1)\right] \tag{1.8-6}$$

or, in terms of the original variables

$$\omega(x,y) = f\left[\omega(x_1,y_1)\right]. \tag{1.8-7}$$

Written inversely, the necessary and sufficient condition for <u>invariance of the family (1.8-1) under the group (1.8-5)</u> is the existence of a function g such that

$$\omega(x_1,y_1) = g\left[\omega(x,y)\right] \quad \text{for all} \quad (x,y). \tag{1.8-8}$$

<u>Example</u>: In the example above of straight lines we had

$$\omega(x_1,y_1) = \omega(x,y) + \text{const.} \qquad \text{for all} \quad x,y.$$

Next, we can apply the idea to all the transformations of a one-parameter group. Represent the group infinitesimally by U,

$$Uf \equiv \xi \frac{\partial f}{\partial x} + \eta \frac{\partial f}{\partial y} \tag{1.8-9}$$

and remember that, locally, the global transformations have the expansion

$$\begin{aligned} x_1 &= x + \tau Ux + \frac{\tau^2}{2!} U^2x + \cdots \\ y_1 &= y + \tau Uy + \frac{\tau^2}{2!} U^2y + \cdots \end{aligned} \tag{1.8-10}$$

Thus, also (cf. 1.5-23)

$$\omega(x_1,y_1) = \omega(x,y) + \tau U\omega + \frac{\tau^2}{2!} U^2\omega + \cdots . \tag{1.8-11}$$

For invariance we want (1.8-8) to hold; a necessary condition is thus that

$$U\omega \equiv \xi \frac{\partial \omega}{\partial x} + \eta \frac{\partial \omega}{\partial y} = \Omega(\omega) \tag{1.8-12}$$

for some function $\Omega(\omega)$. This is also sufficient since

$$U^2\omega = U(U\omega) = U\left[\Omega(\omega)\right] = \frac{d\Omega}{d\omega} U\omega = \Omega \frac{d\Omega}{d\omega} = fn(\omega) .$$

<u>In summary</u>: The family $(\infty^1)\omega(x,y) = \text{const.}$ admits a group U iff

$$U\omega = \Omega(\omega) \quad \text{for some function} \quad \Omega(\omega),$$

when $\Omega \equiv 0$ then each curve is mapped to itself.

<u>Example</u>: The family of rays $\omega(x,y) = y/x = \text{const.}$ is left invariant under rotation

$$Uf = -y \frac{\partial f}{\partial x} + x \frac{\partial f}{\partial y} .$$

Note that

$$U\omega = -y\frac{\partial\omega}{\partial x} + x\frac{\partial\omega}{\partial y} = \frac{y^2}{x^2} + 1 = \omega^2 + 1;$$

each ray is mapped into another; however, for the family of circles

$$\omega = x^2 + y^2 = \text{const.}$$

$U\omega = -y(2x) + x(2y) = 0$; each circle goes into itself.

The converse idea is to find all families of curves which admit a given group. Eventually (1.8-12) must be solved for ω which at first appears ill-defined since $\Omega(\omega)$ is arbitrary. The following example illustrates in detail that we can choose $\Omega = 1$, without loss of generality.

Example:

$$Uf \equiv -y\frac{\partial f}{\partial x} + x\frac{\partial f}{\partial y} \quad \text{rotation}$$

We want

$$U\omega \equiv -y\frac{\partial\omega}{\partial x} + x\frac{\partial\omega}{\partial y} = \Omega(\omega).$$

Note that if $\Omega = 0$; $\omega = x^2 + y^2$ or $\omega = F(x^2+y^2)$, F arbitrary. We can always write any family of curves $\omega(x,y) = \text{const.}$ as $G(\omega(x,y)) = \text{const.}$; then, if $\Omega \neq 0$

$$UG \equiv \frac{dG}{d\omega}U\omega = \frac{dG}{d\omega}\Omega$$

and we can choose G such that $dG/d\omega\ \Omega = 1$; G is found from $UG = 1$. We can replace G by ω and solve

$$U\omega = -y\frac{\partial\omega}{\partial x} + x\frac{\partial\omega}{\partial y} = 1.$$

The characteristic differential equations for this P.D.E. are

$$\frac{dx}{-y} = \frac{dy}{x} = \frac{d\omega}{1}.$$

The integral curves of the first pair are $x^2 + y^2 = r^2 = \text{const.}$ and along such a curve

$$\frac{d\omega}{1} = \frac{dy}{\sqrt{r^2-y^2}}$$

thus: $\omega = \sin^{-1} y/r + f(r)$. The general solution of the PDE containing the arbitrary function f is

$$\omega = \tan^{-1}\frac{y}{x} + f(x^2+y^2).$$

Problem 1.8-1. Given a one-parameter family of ellipses $(\tau = \tau(\sigma))$

$$\frac{x^2}{\tau^2} + \frac{y^2}{\sigma^2} = 1,$$

find a group of transformations leaving these invariant.

Hint: consider the projective group.

1.9. First-Order Differential Equations Which Admit a Group; Integrating Factor; Commutator

In this section we show that invariance of a first-order differential equation under a group leads to the construction of an integrating factor and a reduction to quadrature.

An (∞^1) family of curves $\omega(x,y) = \text{const.}$ can be thought of as the integral curves of a first-order differential equation

$$\frac{dy}{dx} = F(x,y) = \frac{Y(x,y)}{X(x,y)} \tag{1.9-1}$$

or, written in differential form

$$X(x,y)dy - Y(x,y)dx = 0.$$

If $\omega(x,y) = \text{const.}$ are integral curves, then

$$X(x,y)\frac{\partial\omega}{\partial x} + Y(x,y)\frac{\partial\omega}{\partial y} = 0 \quad \text{for all } x,y. \tag{1.9-2}$$

Now assume that the family of integral curves (known or unknown) associated with the differential equation admits the group U. Then (cf. 1.8-12) Ω exists such that

$$U\omega \equiv \xi(x,y)\frac{\partial\omega}{\partial x} + \eta(x,y)\frac{\partial\omega}{\partial y} = \Omega(\omega). \tag{1.9-3}$$

Note the point of working with these representations: all considerations involve points of the (x,y) plane, constants defining curves do not appear. Further, note that for any Φ, $\Phi(\omega) = \text{const.}$ is also a representation of the integral

$$X\frac{\partial\Phi}{\partial x} + Y\frac{\partial\Phi}{\partial y} = \frac{d\Phi}{d\omega}\left\{X\frac{\partial\omega}{\partial x} + Y\frac{\partial\omega}{\partial y}\right\} = 0 \tag{1.9-4}$$

and Φ admits the group

$$U\Phi = \xi\frac{\partial\Phi}{\partial x} + \eta\frac{\partial\Phi}{\partial y} = \frac{d\Phi}{d\omega}U\omega = \Omega(\omega)\frac{d\Phi}{d\omega}. \tag{1.9-5}$$

Now assume

$$\Omega(\omega) \not\equiv 0$$

which means that each integral curve does not go into itself under the transformations. Then it is possible to choose Φ such that the right hand side of (1.9-5) is one.

Therefore, for the family of integral curves of (1.9-1) invariant under the group with infinitesimal U we have (Φ replaced by ω)

$$\begin{aligned} X\frac{\partial\omega}{\partial x} + Y\frac{\partial\omega}{\partial y} &= 0 \qquad \text{d. e.} \\ \xi\frac{\partial\omega}{\partial x} + \eta\frac{\partial\omega}{\partial y} &= 1 \qquad \text{invariance} \end{aligned} \tag{1.9-6}$$

This system (1.9-6) can be solved for the first partial derivatives

$$\frac{\partial\omega}{\partial x} = -\frac{Y}{X\eta - Y\xi}, \quad \frac{\partial\omega}{\partial y} = \frac{X}{X\eta - Y\xi}. \tag{1.9-7}$$

Thus, the first partial derivatives of a representation $\omega(x,y) = \text{const.}$ of the integral curves are known as functions of (x,y); <u>this means $d\omega$ is known exactly</u>

$$d\omega = \frac{\partial\omega}{\partial x}\,dx + \frac{\partial\omega}{\partial y}\,dy = \frac{Xdy - Ydx}{X\eta - Y\xi} . \tag{1.9-8}$$

The construction on the right hand side of (1.9-8) is the differential of a representation $\omega(x,y) = \text{const.}$ of the integral curves; thus it can be integrated and

$$M = \frac{1}{X\eta - Y\xi} \tag{1.9-9}$$

is the integrating factor.(1) If (ξ,η) are known the integration problem for (1.9-1) is reduced to quadrature.

Commutator. We now need to develop a criterion that a given differential equation admits a group U since, in general, the differential equation (1.9-1) is given rather than the integral curves (which are sought). This criterion can be expressed with the help of the commutator $[U,A]$.

Two operators enter the above considerations

$$Uf \equiv \xi\frac{\partial f}{\partial x} + \eta\frac{\partial f}{\partial y} , \quad \text{derivative in direction of transformation} \tag{1.9-10}$$

$$Af \equiv X\frac{\partial f}{\partial x} + Y\frac{\partial f}{\partial y} , \quad \text{derivative in direction of integral curves} \tag{1.9-11}$$

The commutator is also a first-order operator formed from these two

$$\begin{aligned}[U,A]f &\equiv U(Af) - A(Uf) \\ &= \left(\xi\frac{\partial}{\partial x} + \eta\frac{\partial}{\partial y}\right)\left(X\frac{\partial f}{\partial x} + Y\frac{\partial f}{\partial y}\right) \\ &\quad - \left(X\frac{\partial}{\partial x} + Y\frac{\partial}{\partial y}\right)\left(\xi\frac{\partial f}{\partial x} + \eta\frac{\partial f}{\partial y}\right) .\end{aligned} \tag{1.9-12}$$

It is clear that all second derivatives drop out so that

$$[U,A]f \equiv (UX - A\xi)\frac{\partial f}{\partial x} + (UY - A\eta)\frac{\partial f}{\partial y} . \tag{1.9-13}$$

(1) This important result is due to S. Lie, 1874 Verh. Gesell, d. Wissenschaften zu Christiania.

If $\omega(x,y) = \text{const.}$ are integral curves admitting the group U, then from (1.9-2) and (1.9-3)

$$[U,A]\omega \equiv U(A\omega) - A(U\omega) = -A\Omega(\omega) = -\frac{d\Omega}{d\omega} A\omega = 0. \tag{1.9-14}$$

Thus

$$\begin{aligned} [U,A]\omega &= 0 = (UX - A\xi)\frac{\partial\omega}{\partial x} + (UY - A\eta)\frac{\partial\omega}{\partial y} \\ A\omega &= 0 = X\frac{\partial\omega}{\partial x} + Y\frac{\partial\omega}{\partial y}\,. \end{aligned} \tag{1.9-15}$$

Hence there exists a function $\lambda(x,y)$ such that for each (x,y)

$$UX - A\xi = \lambda(x,y)X, \qquad UY - A\eta = \lambda(x,y)Y. \tag{1.9-16}$$

Thus, the operator condition for invariance of a given differential equation is obtained: $\lambda(x,y)$ must exist such that

$$[U,A]f \equiv \lambda\left\{X\frac{\partial f}{\partial x} + Y\frac{\partial f}{\partial y}\right\} = \lambda Af. \tag{1.9-17}$$

The argument can also be reversed.

<u>In summary</u>: The ordinary differential equation $Xdy - Ydx = 0$ admits the one-parameter group defined by U,

$$Uf \equiv \xi\frac{\partial f}{\partial x} + \eta\frac{\partial f}{\partial y}$$

if and only if for

$$Af \equiv X\frac{\partial f}{\partial x} + Y\frac{\partial f}{\partial y}\,,$$

$\lambda(x,y)$ exists such that

$$[U,A]f \equiv U(Af) - A(Uf) = \lambda(x,y)Af.$$

It is interesting to note how the same criterion can be derived from local considerations. Invariance of the family means that integral curves $\omega(x,y) = \text{const.}$ go into integral curves

$\omega(x_1,y_1) = \text{const}$. In terms of the differential equation (1.9-1) this should imply, for invariance

$$X(x_1,y_1)dy_1 - Y(x_1,y_1)dx_1 = 0. \qquad (1.9\text{-}18)$$

But, the transformation locally is

$$x_1 = x + \xi(x,y)\tau, \qquad y_1 = y + \eta(x,y)\tau.$$

Hence the expression

$$\begin{aligned} X(x_1,y_1)dy_1 - Y(x_1,y_1)dx_1 &= X(x+\xi\tau,\ y+\eta\tau)[dy + (\eta_x dx+\eta_y dy)\tau] \\ &\quad - Y(x+\xi\tau,\ y+\eta\tau)[dx + (\xi_x dx+\xi_y dy)\tau] \\ &= [X(x,y) + \tau(\xi X_x+\eta X_y)][dy + (\eta_x dx+\eta_y dy)\tau] \\ &\quad - [Y(x,y) + \tau(\xi Y_x+\eta Y_y)][dx + (\xi_x dx+\xi_y dy)\tau]. \end{aligned}$$

Keeping terms of $O(\tau)$ only,

$$\begin{aligned} X(x_1,y_1)dy_1 - Y(x_1,y_1)dx_1 &= X(x,y)dy - Y(x,y)dx + \tau\{(\xi X_x+\eta X_y+X\eta_y-Y\xi_y)dy \\ &\quad - (\xi Y_x+\eta Y_y-X\eta_x+Y\xi_x)dx\}. \end{aligned} \qquad (1.9\text{-}19)$$

In order that $Xdy - Ydx = 0$ imply $X_1dy_1 - Y_1dx_1 = 0$ we need ξ, η, $\rho(x,y)$ such that

$$\begin{aligned} \xi X_x + \eta X_y + X\eta_y - Y\xi_y &= \rho(x,y)X \\ \xi Y_x + \eta Y_y - X\eta_x + Y\xi_x &= \rho(x,y)Y \end{aligned} \qquad (1.9\text{-}20)$$

and then $X_1dy_1 - Y_1dx_1 = [1 + \tau\rho(x,y)][Xdy - Ydx] = 0$. The conditions (1.9-20) can be brought to the previous form (1.9-16) since

$$\begin{aligned} \rho(x,y) - (\xi_x+\eta_y) &= \frac{UX + X\eta_y - Y\xi_y}{X} - (\xi_x+\eta_y) \\ &= \frac{UY - X\eta_x + Y\xi_x}{Y} - (\xi_x+\eta_y) \\ &= \frac{UX - A\xi}{X} = \frac{UY - A\eta}{Y} = \lambda(x,y). \end{aligned}$$

Thus, the commutator criterion (1.9-17) is obtained.

Example: Differential equation for rays: $xdy - ydx = 0$

derivative along integral curves

$$Af = x\frac{\partial f}{\partial x} + y\frac{\partial f}{\partial y}$$

rotation group

$$Uf = -y\frac{\partial f}{\partial x} + x\frac{\partial f}{\partial y}$$

commutator

$$[U,A]f = \left(-y\frac{\partial}{\partial x} + x\frac{\partial}{\partial y}\right)\left(x\frac{\partial f}{\partial x} + y\frac{\partial f}{\partial y}\right)$$

$$- \left(x\frac{\partial}{\partial x} + y\frac{\partial}{\partial y}\right)\left(-y\frac{\partial f}{\partial x} + x\frac{\partial f}{\partial y}\right)$$

$$= -y\frac{\partial f}{\partial x} + x\frac{\partial f}{\partial y} - x\frac{\partial f}{\partial y} + y\frac{\partial f}{\partial x} \equiv 0.$$

Note: since the commutator $[U,A]$ is identically zero, the roles of U,A can be interchanged and

differential equation $-y\,dy \quad -x\,dx = 0$ (circles)

admits $Uf = x\frac{\partial f}{\partial x} + y\frac{\partial f}{\partial y}$ (stretching group)

Example: The family of circles tangent to the (x,y) axes is invariant under the stretching group - hence find the orthogonal trajectories. Method: The differential equation for orthogonal trajectories admits the same group so that an integrating factor can be found.

Another general method of integration when the group U is known is the use of canonical coordinates. That is (r,s) are introduced so that the group reads

$$Uf \equiv \frac{\partial f}{\partial s}\,. \tag{1.9-21}$$

After this is done quadrature is merely a <u>separation of variables</u>. The canonical coordinates are found by a simple quadrature when the path curves of the group are known (cf. 1.5-20, 1.5-21).

path curves $r = r(x,y) = \text{const.}$

choose $s(x,y)$ such that $Us = \xi s_x + \eta s_y = 1.$

Expressed in (r,s) the original differential equation (1.9-1) has the form

$$ds - F(r,s)dr = 0 \tag{1.9-22}$$

but this admits the group U given by (1.9-21). Since the group is now one of translation $(r_1 = r,\ s_1 = s + \alpha)$, $F(r,s)$ cannot contain s. A formal proof follows from the commutator condition for invariance (1.9-17)

$$Uf = \frac{\partial f}{\partial s}, \qquad Af = \frac{\partial f}{\partial r} + F(r,s)\frac{\partial f}{\partial s}$$

and

$$[U,A]f = \frac{\partial}{\partial s}\left(\frac{\partial f}{\partial r} + F\frac{\partial f}{\partial s}\right) - \left(\frac{\partial}{\partial r} + F\frac{\partial}{\partial s}\right)\frac{\partial f}{\partial s} = \frac{\partial F}{\partial s}\frac{\partial f}{\partial s}$$

must be identically equal to

$$\lambda(r,s)\left\{\frac{\partial f}{\partial r} + F\frac{\partial f}{\partial s}\right\}. \tag{1.9-23}$$

For arbitrary f this can only be true when $\lambda = 0$ and

$$\frac{\partial F}{\partial s} = 0, \qquad F = F(r). \tag{1.9-24}$$

Thus, in (s,r) coordinates the solution is a quadrature

$$s = \int F(r)dr + \text{const.} \tag{1.9-25}$$

<u>Note</u>: if a slightly more general $s(x,y)$ is chosen such that

$$Us = G(s)$$

then the argument used above shows that

$$F(r,s) = \frac{G(s)}{H(r)}$$

and the differential equation reads

$$\frac{ds}{G(s)} = \frac{dr}{H(r)}$$

again a quadrature by separation of variables.

Example: The homogeneous equation $\frac{dy}{dx} = F(\frac{y}{x})$ is obviously invariant under the stretching group $(x_1 = \alpha x,\ y_1 = \alpha y)$

$$Uf = x\frac{\partial f}{\partial x} + y\frac{\partial f}{\partial y}$$

$$Af = \frac{\partial f}{\partial x} + F(\tfrac{y}{x})\frac{\partial f}{\partial y}$$

$$[U,A]f = \left(x\frac{\partial}{\partial x} + y\frac{\partial}{\partial y}\right)\left(\frac{\partial f}{\partial x} + F(\tfrac{y}{x})\frac{\partial f}{\partial y}\right) - \left(\frac{\partial}{\partial x} + F(\tfrac{y}{x})\frac{\partial}{\partial y}\right)\left(x\frac{\partial f}{\partial x} + y\frac{\partial f}{\partial y}\right)$$

$$= -\frac{\partial f}{\partial x} - F(\tfrac{y}{x})\frac{\partial f}{\partial y} = -Af; \quad \lambda = -1$$

Canonical coordinates: path curves $r = \frac{y}{x}$

$$s = \log x, \text{ so } xs_x + ys_y = 1$$

then

$$y = e^s r \qquad dy = e^s dr + re^s ds$$

and

$$\frac{dr}{ds} + r = F(r) \quad \text{or} \quad \frac{dr}{F(r)-r} = ds.$$

Problem 1.9-1. Show that if a first-order differential equation admits two nontrivial groups U_1, U_2 then either

$$\Omega = \frac{X\eta_2 - Y\xi_2}{X\eta_1 - Y\xi_1}$$

is an integral or simply a constant.

1.10. Geometric Interpretation of the Integrating Factor

If $\underline{\xi} = (\xi(x,y), \eta(x,y))$ is the infinitesimal of the group leaving invariant the integral curves $\omega(x,y) = \text{const.} = c$ of $Xdy - Ydx = 0$, then the integrating factor, $M(x,y) = \frac{1}{X\eta - Y\xi}$, evaluated at $\underline{r} = (x,y)$, is inversely proportional to the area of the parallelogram formed by the vectors $\underline{X} = (X(x,y, Y(x,y))$ (which is tangent to the integral curve at $\underline{r}$), and $\underline{\xi}$ (which is tangent to the path curve of the group at $\underline{r}$)

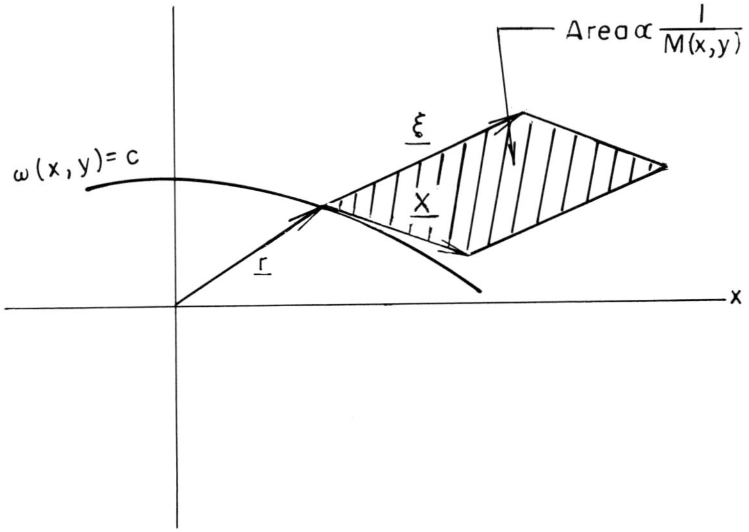

Figure 1.10-1

Another way of looking at the integrating factor geometrically is to consider neighboring integral curves $\omega(x,y) = c$, $\omega(x',y') = c + \delta c$. Let $\delta \underline{r} = (\delta x, \delta y)$ be normal to $\omega(x,y) = c$ at $\underline{r}$, where $\delta s = |\delta \underline{r}|$ is the distance from $\omega = c$ to $\omega = c + \delta c$ at $\underline{r}$.

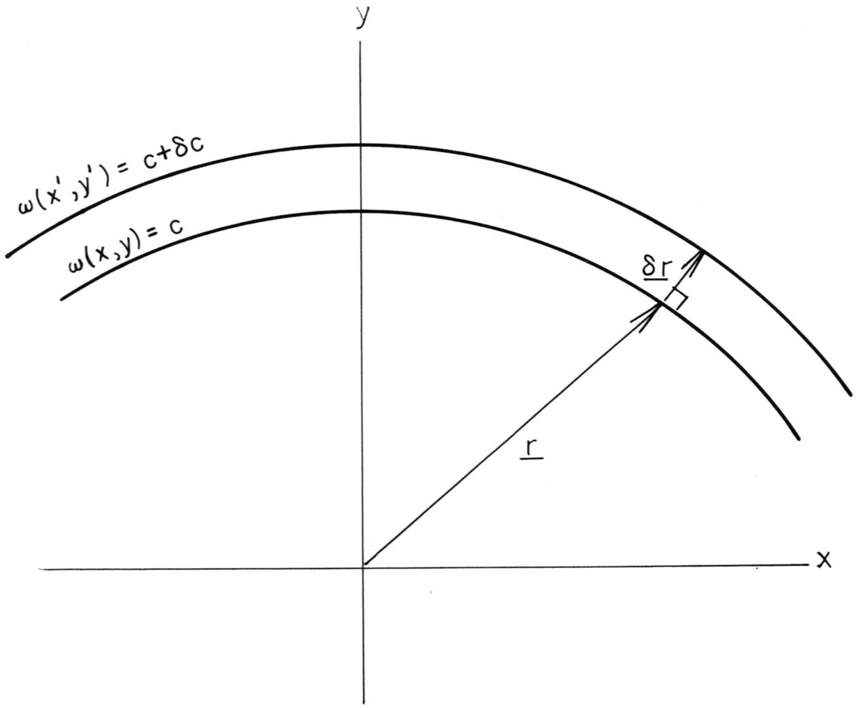

Figure 1-10-2

Then along $\delta \underline{r}$,

$$\delta\omega = \delta c = \frac{\partial\omega}{\partial x}\,\delta x + \frac{\partial\omega}{\partial y}\,\delta y. \tag{1.10-1}$$

Since $\omega(x,y) = \text{const.}$ describes the integral curves obtained from the integrating factor M, $\partial\omega/\partial x = -MY$, $\partial\omega/\partial y = MX$, and hence

$$\sqrt{\left(\frac{\partial\omega}{\partial x}\right)^2 + \left(\frac{\partial\omega}{\partial y}\right)^2} = M\sqrt{X^2+Y^2}\ . \tag{1.10-2}$$

Because $\delta \underline{r}$ and $\nabla\omega = (\partial\omega/\partial x, \partial\omega/\partial y)$ are both orthogonal to $\omega(x,y) = c$, for each $\underline{r}$ on this curve, we must have

$$\delta \underline{r} = \lambda(x,y)\nabla\omega. \tag{1.10-3}$$

Then using (1.10-1), (1.10-3), we find that

$$\frac{\sqrt{\delta x^2 + \delta y^2}}{\sqrt{\left(\frac{\partial\omega}{\partial x}\right)^2 + \left(\frac{\partial\omega}{\partial y}\right)^2}} = \frac{\frac{\partial\omega}{\partial x}\,\delta x + \frac{\partial\omega}{\partial y}\,\delta y}{\left(\frac{\partial\omega}{\partial x}\right)^2 + \left(\frac{\partial\omega}{\partial y}\right)^2}$$

$$\frac{\delta s}{M\sqrt{X^2+Y^2}} = \frac{\delta c}{M^2(X^2+Y^2)}$$

and thus

$$M = \frac{\delta c}{\delta s}\ \frac{1}{\sqrt{X^2+Y^2}}\ . \tag{1.10-4}$$

From this result, it is easily seen that $1/\sqrt{X^2+Y^2}$ is an integrating factor if it is known that the integral curves are parallel to each other. Next we will derive this latter result from the group point of view.

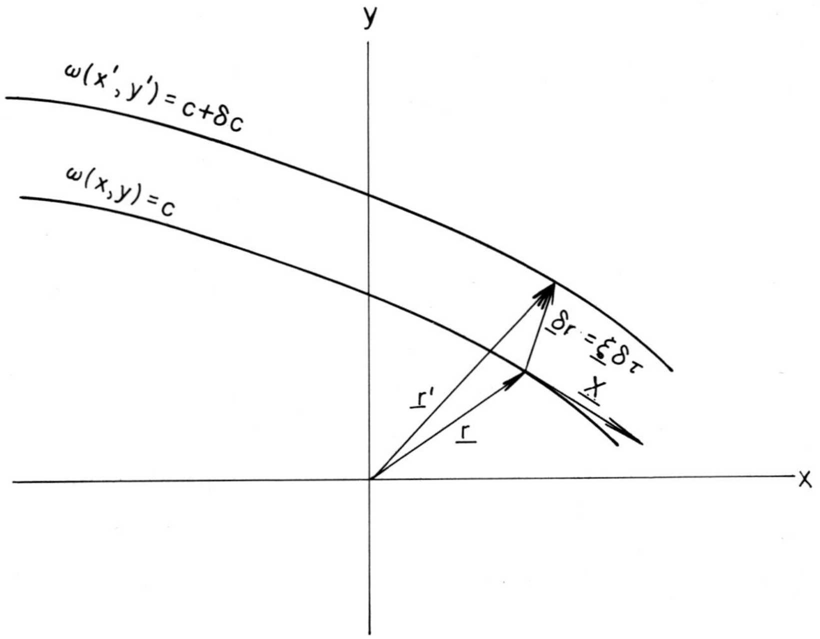

Figure 1.10-3

Since $\underline{\xi}$ gives the "velocity" for going from one integral curve to another, in the case of parallel integral curves we must have $\underline{\xi} \perp \underline{X} \Rightarrow X\xi = -\eta Y$. Also in this case, if $\underline{r}' = (x',y') = \underline{x} + \delta\tau\underline{\xi}$ lies on $\omega = c + \delta c$, then $|\delta\underline{r}| = |\underline{r}'-\underline{r}| = \text{const.}$ for all $\underline{r}$ lying on $\omega = c$. Hence $\sqrt{\xi^2+\eta^2} = \text{const.}$

$$\sqrt{\xi^2+\eta^2} = \sqrt{\frac{\eta^2Y^2}{X^2} + \eta^2} = \frac{\eta}{X}\sqrt{Y^2+X^2}$$

$$\frac{\eta}{X} = \frac{\text{const.}}{\sqrt{Y^2+X^2}}$$

$$\therefore \quad \frac{1}{X\eta - Y\xi} = \frac{\eta}{X}\,\frac{1}{\eta^2 - \frac{Y\xi\eta}{X}} = \frac{\eta}{X} \cdot \frac{1}{\eta^2 + \xi^2} = \frac{\text{const.}}{\sqrt{Y^2+X^2}}$$

The result shows that for parallel integral curves $\operatorname{div} \vec{t} \equiv 0$ where $\vec{t}$ is the unit tangent to an integral curve.

1.11. Determination of First-Order Equations Which Admit a Given Group

From the fact that the group is known the general form of a first-order differential equation which is invariant and so integrable by quadrature can be found. The converse, and possibly more interesting problem, of finding a group (there may be several) for a given first-order differential equation is more difficult. No systematic method exists; geometric intuition or trial and error involving equations of the infinitesimals (ξ, η) which are developed later (§1.13) must be used.

The differential equation defines a family (∞^1) of integral curves. However, any curve in (x,y) which is not a path curve of the group generates another curve under transformation by a member of a group of transformations. Under a one-parameter group any curve is equivalent to an (∞^1) family of curves corresponding to the given curve. The totality (∞^1) of these curves are equivalent by the basic group property. If we think of this (∞^1) family as the integral curves of a differential equation then the differential equation can be found by differentiation and elimination of the parameter defining a particular curve.

Example (1): translation along x $\quad x_1 = x + \alpha$, $\quad y_1 = y$.

Take any curve and note two possibilities

(i) the equation of the curve does not contain x: $y = \text{const.}$ The differential equation of this family is $dy/dx = 0$; these are path curves.

(ii) the curve can be written as $x - f(y) = 0$; under translation we have

$$x_1 - f(y_1) = 0 = x + \alpha - f(y).$$

Thus the family of curves is

$$x - f(y) = \text{const.}$$

Taking

$$\frac{d}{dx}: \ dx - f'(y)dy = 0$$

or the differential equation admitting the translation group with respect to x is

$$\frac{dy}{dx} = g(y)$$

where g is an arbitrary function of y.

Example (2): Affine group

$$x_1 = \alpha x$$
$$y_1 = y$$

(i) $y = \text{const.}\ \frac{dy}{dx} = 0$

(ii) $x_1 - f(y_1) = 0$

or

$$\alpha x - f(y) = 0$$

thus

$$\frac{f(y)}{x} = \text{const.}$$

$$\frac{f'(y)dy}{x} - \frac{f(y)}{x^2}dx = 0$$

or D.E. admitting affine group is

$$xdy - F(y)dx = 0$$

where F is an arbitrary function of y. Note: group has $Uf = x\frac{\partial f}{\partial x}$ integrating factor:

$$M = \frac{1}{X\eta - Y\xi} = \frac{1}{-xF(y)} .$$

Thus separation of variables can be used.

Example (3): Stretching or Perspective Group, $x_1 = \alpha x \quad y_1 = \alpha y$. If $x_1 - f(y_1) = 0$ is a member of the family then $\alpha x - f(\alpha y) = 0$ are all the curves of the family or $x = 1/\alpha \ f(\alpha y)$ defines the family

$$dx = f'(\alpha y)dy$$

or solving this

$$\alpha y = g(y') .$$

So

$$x = \frac{1}{\alpha} f(g(y'))$$

or

$$\frac{x}{y} = \frac{1}{\alpha y} f(g(y')) = \frac{f(g(y'))}{g(y')}$$

Finally, solving this equation we see that

$$y' = F(\tfrac{y}{x})$$

admits the persepctive group where F is an arbitrary function of $\frac{y}{x}$.

Example (4): Rotation Group

Show that the corresponding differential equation is:

$$\frac{xy' - y}{x + yy'} = F(x^2+y^2)$$

where F is an arbitrary function of $x^2 + y^2$.

Example (5): Group for a Linear Differential Equation

$$x_1 = x$$

$$y_1 = y + \alpha\phi(x)$$

if

$$y_1 - f(x_1) = 0 \quad \text{is a curve}$$

then

$y + \alpha\phi(x) - f(x) = 0$ defines the one-parameter family of curves

$$\frac{dy}{dx} = f'(x) - \alpha\phi'(x) = f'(x) - \phi'(x)\,\frac{f(x) - y}{\phi(x)}$$

or

$$\frac{dy}{dx} - \Phi(x)y - F(x) = 0 \quad \text{where} \quad \Phi(x) = \frac{\phi'(x)}{\phi(x)}$$

and $F(x)$ is an arbitrary function of x. Note: this equation admits the group

$$x_1 = x$$

$$y_1 = y + \alpha e^{\int \Phi dx}$$

$$Uf = e^{\int \Phi(x)dx}\,\frac{\partial f}{\partial y}$$

and the integrating factor is the usual one

$$M = \frac{1}{X\eta - Y\xi} = e^{-\int \Phi(x)dx}.$$

1.12. One-Parameter Group in Three Variables; More Variables

The ideas of the previous sections can be extended in various ways; more variables, more parameters. For a study of higher order differential equations it is essential to talk of more variables. All the basic ideas already appear in the case of three variables (x,y,z) for the extension to n variables.

The primary application for the case of three variables is to the system

$$\frac{dy}{dx} = \frac{Y(x,y,z)}{X(x,y,z)}, \qquad \frac{dz}{dx} = \frac{Z(x,y,z)}{X(x,y,z)} \qquad (1.12\text{-}1)$$

or simply

$$\frac{dx}{X(x,y,z)} = \frac{dy}{Y(x,y,z)} = \frac{dz}{Z(x,y,z)} . \tag{1.12-2}$$

As usual, (1.12-2) defines the local direction field of an integral curve at each (regular) point in (x,y,z) space (Fig. 1.12-1). Integration consists in finding all the integral curves, a doubly infinite (∞^2) family. One integral curve passes through each (non-singular) point of the space. The system of equations (1.12-2) is connected to the first order P.D.E.

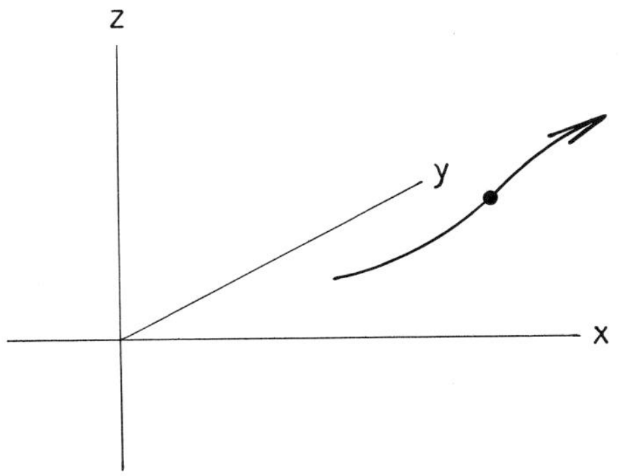

Figure 1.12-1

$$X \frac{\partial f}{\partial x} + Y \frac{\partial f}{\partial y} + Z \frac{\partial f}{\partial z} = 0. \tag{1.12-3}$$

The curves defined by (1.12-2) are, of course, the characteristic curves of the linear P.D.E. (1.12-3).

The integral curves of (1.12-3) can be represented by two functions u,v with

$$\begin{aligned} u(x,y,z) &= a = \text{const.} \\ v(x,y,z) &= b = \text{const.} \end{aligned} . \tag{1.12-4}$$

A particular pair of values (a,b) defines one curve in (x,y,z) space which is the intersection of the surfaces $(u = a,\ v = b)$. Each surface $u = \text{const.}$ or $v = \text{const.}$ is thus swept out by a one-parameter family of integral curves. Therefore, as a definition:

Integral: $u(x,y,z)$ is an integral of (1.12-1 and 1.12-2) when each surface $u(x,y,z) = \text{const.}$ is generated by a one-parameter family (∞^1) of integral curves. This occurs if, at each point of u

$$X \frac{\partial u}{\partial x} + Y \frac{\partial u}{\partial y} + Z \frac{\partial u}{\partial z} = 0. \tag{1.12-5}$$

If two mutually independent integrals $u(x,y,z)$, $v(x,y,z)$ are known then

$$\Omega(u,v) = 0 \qquad \Omega \text{ arbitrary} \tag{1.12-6}$$

is the general equation of a surface swept out by a one-parameter family of integral curves (∞^1) and so is the general integral of (1.12-2), i.e., $u = a$, $v = b(a)$ is one curve.

Any single surface

$$F(x,y,z) = 0 \tag{1.12-7}$$

is generated by (∞^1) integral curves if at each point the integral curve has the tangent direction to the surface, that is

$$X \frac{\partial F}{\partial x} + Y \frac{\partial F}{\partial y} + Z \frac{\partial F}{\partial z} = 0 \quad \text{when} \quad F = 0. \tag{1.12-8}$$

Thus, $F(x,y,z) = 0$ can be written as $\Omega(u,v) = 0$ when u,v are two independent integrals and

$$f = \Omega(u,v) \tag{1.12-9}$$

is the general solution of the associated P.D.E.

$$X \frac{\partial f}{\partial x} + Y \frac{\partial f}{\partial y} + Z \frac{\partial f}{\partial z} = 0. \tag{1.12-10}$$

Examples:

(i)

$$\frac{dx}{y} = \frac{dy}{-x} = \frac{dz}{0}$$

$$y \frac{\partial f}{\partial x} - x \frac{\partial f}{\partial y} = 0$$

$$u = x^2 + y^2,\ v = z \qquad f = z - F(x^2+y^2) = \Omega(z,x^2+y^2).$$

The integral curves or characteristics are circles $x^2 + y^2 = \text{const.}$ The surface of revolution $x^2 + y^2 = a$ is generated by an (∞^1) of integral curves (different $z = v - b$). The plane $z = b$ is generated by the one-parameter family of integral curves $x^2 + y^2 = = a$ (different a). The general surface swept out by such curves is $\Omega(z, x^2+y^2) = 0$, or $z = F(x^2+y^2)$ represents an arbitrary surface of revolution. $f = z - F(x^2+y^2)$ is the general solution of the associated linear P.D.E.

(ii) $$\frac{dx}{x} = \frac{dy}{y} = \frac{dz}{z}$$

$$x\frac{\partial f}{\partial x} + y\frac{\partial f}{\partial y} + z\frac{\partial f}{\partial z} = 0.$$

Integral curves are rays through the origin specified, for example by two-angles; any canonical surface generated by rays is an integral surface; the reader can work out the details.

One-Parameter Group in Space

Transformations of the (x,y,z) space into itself take points into points, lines into lines, surfaces into surfaces. As before, the one-parameter family

$$\begin{aligned} x_1 &= \phi(x,y,z;\ \alpha) \\ y_1 &= \psi(x,y,z;\ \alpha) \\ z_1 &= \chi(x,y,z;\ \alpha) \end{aligned} \tag{1.12-11}$$

forms a group of transformations when the group properties hold, as in two dimensions, in particular the composition formula. The existence of the inverse to α is assumed and thus the identity element α_o exists. The infinitesimal transformation describes the neighborhood of the identity or

$$x_1 = x + \left(\frac{\partial\phi}{\partial\alpha}\right)_{\alpha_0} \delta\alpha$$

$$y_1 = y + \left(\frac{\partial\psi}{\partial\alpha}\right)_{\alpha_0} \delta\alpha$$

$$z_1 = z + \left(\frac{\partial\chi}{\partial\alpha}\right)_{\alpha_0} \delta\alpha$$

if these derivatives exist and are not zero. We may proceed to higher terms if necessary but, in general, the infinitesimal transformation exists

$$\begin{aligned} x_1 &= x + \xi(x,y,z)\,\delta\tau \\ y_1 &= y + \eta(x,y,z)\,\delta\tau \\ z_1 &= z + \zeta(x,y,z)\,\delta\tau \end{aligned} \quad . \tag{1.12-12}$$

Example: The screw transformation:

$$(\alpha \to \delta\alpha)$$

$$\begin{aligned} x_1 &= x\cos\alpha - y\sin\alpha \to x - y\delta\alpha, & \xi &= -y \\ y_1 &= y\cos\alpha + x\sin\alpha \to y + x\delta\alpha, & \eta &= x \\ z_1 &= z + m\alpha \to z + m\delta\alpha, & \zeta &= m \end{aligned}$$

The transformation is rotation about the z-axis and translation in the z-direction.

The one-parameter group of transformations is constructed from the infinitesimal transformation by integration of

$$\frac{dx_1}{\xi(x_1,y_1,z_1)} = \frac{dy_1}{\eta(x_1,y_1,z_1)} = \frac{dz_1}{\zeta(x_1,y_1,z_1)} = d\tau \tag{1.12-13}$$

with the initial conditions at $\tau = 0$: $x_1 = x$, $y_1 = y$, $z_1 = z$. The global integrals

$$x_1 = \phi(x,y,z;\ \tau)$$
$$y_1 = \psi(x,y,z;\ \tau)$$
$$z_1 = \chi(x,y,z;\ \tau)$$

represent a one-parameter group of transformations. Proof - as in two dimensional case (cf. §1.3, Eq. 1.3-8 ff). The system (1.12-13) can be integrated by a power series in τ

$$\begin{aligned} x_1 &= x + \xi(x,y,z)\tau + (\xi\xi_x+\eta\xi_y+\zeta\xi_z)\ \frac{\tau^2}{2!} + \cdots \\ y_1 &= y = \eta(x,y,z)\tau + (\xi\eta_x+ \cdots\quad)\ \frac{\tau^2}{2!} + \cdots \\ z_1 &= z + \zeta(x,y,z)\tau + \cdots \end{aligned} \tag{1.12-14}$$

The power series is represented simply by the symbol of the infinitesimal transformation, representing the directional derivative in space

$$Uf \equiv \xi \frac{\partial f}{\partial x} + \eta \frac{\partial f}{\partial y} + \zeta \frac{\partial f}{\partial z}\ . \tag{1.12-15}$$

<u>New coordinates</u>: The transformation of the operator U to new coordinates $(\tilde{x},\tilde{y},\tilde{z})$

$$\tilde{x} = \tilde{x}(x,y,z),\quad \tilde{y} = \tilde{y}(x,y,z),\quad \tilde{z} = \tilde{z}(x,y,z) \tag{1.12-16}$$

is carried out as before (cf. 1.5-14)

$$\tilde{U}f \equiv (U\tilde{x})\ \frac{\partial f}{\partial \tilde{x}} + (U\tilde{y})\ \frac{\partial f}{\partial \tilde{y}} + (U\tilde{z})\ \frac{\partial f}{\partial \tilde{z}}\ . \tag{1.12-17}$$

<u>Canonical variables</u>: The transformation can be represented as a translation in (r,s,t) by finding (r,s,t) such that (cf. 1.5-18).

$$Ur = \xi \frac{\partial r}{\partial x} + \eta \frac{\partial r}{\partial y} + \zeta \frac{\partial r}{\partial z} = 0$$

$$Us = 0 \qquad (1.12\text{-}18)$$

$$Ut = 1$$

Example: Infinitesimal screw transformation

$$Uf = -y \frac{\partial f}{\partial x} + x \frac{\partial f}{\partial y} + m \frac{\partial f}{\partial z}$$

$$\xi = -y, \quad \eta = x, \quad \zeta = m.$$

The canonical coordinates (r,s,t) satisfy

$$-y \frac{\partial r}{\partial x} + x \frac{\partial r}{\partial y} + m \frac{\partial r}{\partial z} = 0$$

$$-y \frac{\partial s}{\partial x} + x \frac{\partial s}{\partial y} + m \frac{\partial s}{\partial z} = 0$$

$$-y \frac{\partial t}{\partial x} + x \frac{\partial t}{\partial y} + m \frac{\partial t}{\partial z} = 1$$

with corresponding characteristic differential equations:

$$\frac{dx}{-y} = \frac{dy}{x} = \frac{dz}{m} = \frac{dr}{0} = \frac{ds}{0} = \frac{dt}{1}$$

$$r = \sqrt{x^2+y^2} = \text{const. on circles}$$

$$s = \text{const. on } \frac{dz}{m} = \frac{dy}{\sqrt{r^2-y^2}}$$

$$s = \frac{z}{m} - \theta; \quad \theta = \tan^{-1}\frac{y}{x} = \sin^{-1}\frac{y}{r}$$

$$t = \frac{z}{m}\,.$$

In these coordinates

$$Uf = \frac{\partial f}{\partial t}\,.$$

Power Series of a Function of $g(x,y,z)$ Along a Path: as before (cf. 1.5-23)

$$g(x_1,y_1,z_1) = g(x,y,z) + \tau Ug + \frac{\tau^2}{2!} U^2 g + \cdots . \qquad (1.12\text{-}19)$$

We apply (1.12-19) to (x,y,z) to obtain a power series representation for the global equation of the group (cf. 1.5-25) and (cf. 1.12-14)

$$\begin{aligned} x_1 &= x + \tau Ux + \frac{\tau^2}{2!} U^2 x + \cdots \\ y_1 &= y + \tau Uy + \frac{\tau^2}{2!} U^2 y + \cdots \\ z_1 &= z + \tau Uz + \frac{\tau^2}{2!} U^2 z + \cdots \end{aligned} \quad . \qquad (1.12\text{-}20)$$

Invariants: A function $\Omega(x,y,z)$ defined in (x,y,z) space is invariant when

$$\Omega(x_1,y_1,z_1) = \Omega(x,y,z) \qquad (1.12\text{-}21)$$

From (1.12-19) the condition for invariance is

$$U\Omega(x,y,z) = 0 \quad \text{for all} \quad x,y,z. \qquad (1.12\text{-}22)$$

Example: Screw transformation

$$U\Omega = -y \frac{\partial\Omega}{\partial x} + x \frac{\partial\Omega}{\partial y} + m \frac{\partial\Omega}{\partial z} = 0$$

two independent invariants are $u = \sqrt{x^2+y^2}$, $v = \theta - \frac{z}{m}$, $\theta = \tan^{-1} \frac{y}{x}$. The general invariant function under this group is

$$\Omega(x,y,z) = F\left(\sqrt{x^2+y^2},\ \theta - \frac{z}{m}\right) .$$

Problem 1.12-1. Find the general invariant for

$$Uf \equiv z \frac{\partial f}{\partial x} + z \frac{\partial f}{\partial y} + \frac{\partial f}{\partial z} .$$

Path curves: To each point (x,y,z) is attached a path curve which is generated when all transformations of the one-parameter group are applied. The (∞^2) path curves are identical with the characteristic curves of the associated P.D.E.

$$Uf = \xi \frac{\partial f}{\partial x} + \eta \frac{\partial f}{\partial y} + \zeta \frac{\partial f}{\partial z} = 0 \qquad (1.12\text{-}23)$$

They are found as integrals of the characteristic D.E.

$$\frac{dx}{\xi(x,y,z)} = \frac{dy}{\eta(x,y,z)} = \frac{dz}{\zeta(x,y,z)} \ . \qquad (1.12\text{-}24)$$

Thus the (∞^2) path curves are represented by

$$\begin{aligned} &\Omega_1(x,y,z) = \text{const.} && \text{when } \Omega_1,\Omega_2 \\ &\Omega_2(x,y,z) = \text{const.} && \text{are invariants} \end{aligned} \qquad (1.12\text{-}25)$$

Example: screw transformation

$$Uf \equiv -y \frac{\partial f}{\partial x} + x \frac{\partial f}{\partial y} + m \frac{\partial f}{\partial z}$$

$$\Omega_1(x,y,z) = u = \sqrt{x^2+y^2}$$

$$\Omega_2(x,y,z) = v = \theta - \frac{z}{m}$$

check:

$$\text{Globally} \begin{cases} x_1 = x\cos\tau - y\sin\tau \\ y_1 = y\cos\tau + x\sin\tau \\ z_1 = z + m\tau \end{cases}$$

so

$$x_1^2 + y_1^2 = x^2 + y^2$$

$$\frac{y_1}{x_1} = \frac{y/x + \tan\tau}{1 - y/x \tan\tau} = \tan(\theta+\tau); \quad \theta_1 = \theta + \tau$$

$$\theta_1 - \frac{z_1}{m} = \theta + \tau - \frac{z + m\tau}{m} = \theta - \frac{z}{m} .$$

The path curves $\left\{x_1(\tau),\ y_1(\tau),\ z_1(\tau)\right\}$ are given by the global transformation and are circular helices.

Invariant Curves and Surfaces of a One-Parameter Group

Invariant curves: These are curves in space which transform into themselves under all members of the group. They are evidently of two kinds

(i) path curves

(ii) curves composed of invariant points: these are defined by $\xi(x,y,z) = 0$, $\eta(x,y,z) = 0$, $\zeta(x,y,z) = 0$. These last three relations may define a surface in which case any curve drawn on the surface is invariant.

Invariant surfaces: These are surfaces in space which transform into themselves under all members of the group; again there are two kinds

(i) surfaces generated by one-parameter families of (∞^1) path curves

$$\Omega(x,y,z) = c \quad \text{is a surface} \tag{1.12-26}$$

then if $\Omega(x_1,y_1,z_1) = c$ the surface is invariant that is

$$\Omega(x,y,z) = \Omega(x_1,y_1,z_1) . \tag{1.12-27}$$

Thus invariant functions define invariant surfaces; in

particular if $u(x,y,z)$, $v(x,y,z)$ are two independent invariants,

$$\Omega = W(u,v) = \text{const.}$$

is the general invariant surface. Analytically:

$$\left.\begin{array}{l}\omega(x,y,z) = 0 \text{ is an invariant surface if} \\ U\omega = 0 \text{ when } \omega = 0 \text{ (not all } \omega_x, \omega_y, \omega_z = \text{zero)}\end{array}\right\} . \qquad (1.12\text{-}28)$$

(ii) surfaces composed of invariant points $(\xi,\eta,\zeta) = 0$

$$U\omega = 0.$$

Problem 1.12-2. Geometric Example: The general projective transformation of the (x,y,z) space which takes planes into planes is

$$Uf \equiv \alpha x \frac{\partial f}{\partial x} + \beta y \frac{\partial f}{\partial y} + \gamma z \frac{\partial f}{\partial z}$$

(i) find path curves

(ii) find invariant surfaces

$$z = x^{\gamma/\alpha} G\left(\frac{y^{\alpha}}{x^{\beta}}\right)$$

(iii) show that the equations of the principal tangent curves to the invariant surface can be found by quadrature.

1.13. Extended Transformation in the Plane

In this section another method is introduced for studying the invariance properties of differential equations. For first-order equations this consists of studying the transformations in the three-dimensional space of (x,y,y') where y' is the slope of a given curve at (x,y). By this extension of transformations in the plane invariance for first-order differential equations can be studied. The idea generalizes easily to second and higher order equations by studying spaces of more dimensions (x,y,y',y''), (x,y,y',y'',y'''), etc.

Details will now be worked out. (x,y,y') are considered to be the coordinates of a three-dimensional space. However, y' is the slope of a given curve (i.e., a direction attached to a point in the (x,y) plane). Hence a knowledge of how the plane transforms should enable us to calculate how y' transforms. (In order to emphasize the three-dimensional nature of the transformation sometimes y' is denoted by p and the (x,y,p) space is considered.) Now let

$$\begin{aligned} x_1 &= \phi(x,y) \\ y_1 &= \psi(x,y) \end{aligned} \tag{1.13-1}$$

be a transformation of the plane into itself and let

$$y = F(x) \tag{1.13-2}$$

be an arbitrary curve.

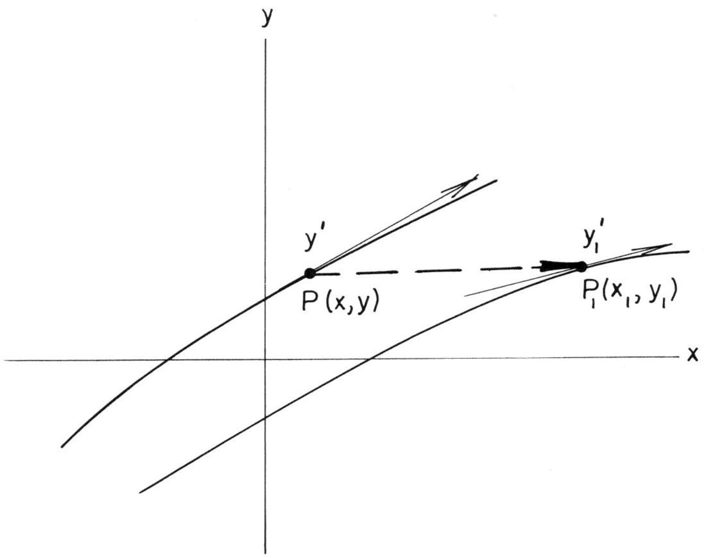

Figure 1.13-1

Then under (1.13-1) a new curve

$$y_1 = F_1(x_1) \tag{1.13-3}$$

is generated. The tangential directions to this curve transform at each point (x,y) as follows

$$y' \equiv \frac{dy}{dx} = F'(x) \tag{1.13-4}$$

$$y_1' \equiv \frac{dy_1}{dx_1} = F_1'(x_1) = \frac{d\psi}{d\phi} = \frac{\psi_x dx + \psi_y dy}{\phi_x dx + \phi_y dy} = \frac{\psi_x + \psi_y y'}{\phi_x + \phi_y y'} . \tag{1.13-5}$$

Thus, we have the extended transformation completely defined in terms of ϕ, ψ

$$\left.\begin{aligned} x_1 &= \phi(x,y) \\ y_1 &= \psi(x,y) \\ y_1' &= \frac{\psi_x + \psi_y y'}{\phi_x + \phi_y y'} \end{aligned}\right\} \quad \text{once extended transformation.} \tag{1.13-6}$$

Evidently, y_1' is the slope of the transformed curve at the point (x_1,y_1) which is the image of (x,y).

Examples: The formulas (1.13-6) are easily worked out for the special cases

(i) Translation

(ii) Rotation

(iii) Affinity $x_1 = \alpha x, \quad y_1 = y, \quad y_1' = \frac{y'}{\alpha}$

(iv) Uniform Stretching

Group Properties

If instead of the single transformation (1.13-1) a one-parameter group is considered, it is clear from geometric considerations that the extended formulas also form a group (each curve $y_1 = F_1(x_1)$

corresponds to a definite parameter value α, composition rules hold for y', the inverse and identity exist). Thus

$$\begin{array}{lll} \text{once} & x_1 = \phi(x,y;\ \alpha) & \\ \text{extended} & y_1 = \psi(x,y;\ \alpha) & (1.13\text{-}7) \\ \text{group} & y_1' = \dfrac{\psi_x(x,y;\ \alpha) + \psi_y(x,y;\ \alpha)y'}{\phi_x(x,y;\ \alpha) + \phi_y(x,y'\alpha)y'} & \end{array}$$

Problem 1.13-1. Check directly for the rotation group.

Infinitesimal Transformation

Thus the infinitesimal version of (1.13-7) can be constructed, which should prove useful for deriving invariance properties of differential equations.

Infinitesimally, in the plane, (1.13-7) is represented by

$$Uf \equiv \xi \frac{\partial f}{\partial x} + \eta \frac{\partial f}{\partial y} \qquad (1.13\text{-}8)$$

corresponding to the local transformation

$$\begin{aligned} x_1 &= x + \xi(x,y)\tau \\ y_1 &= y + \eta(x,y)\tau \end{aligned} \quad . \qquad (1.13\text{-}9)$$

Then

$$y_1' \equiv \frac{dy_1}{dx_1} = \frac{dy + \tau \cdot d\eta}{dx + \tau \cdot d\xi} . \qquad (1.13\text{-}10)$$

In (1.13-10) the changes $(d\eta, d\xi)$ are the changes in (ξ,η) as we travel in the direction y' at the original point (x,y).

$$\frac{d\xi}{dx} \equiv \xi_x + \xi_y y' \quad . \qquad \frac{d\eta}{dx} \equiv \eta_x + \eta_y y' \tag{1.13-11}$$

From (1.13-10)

$$y_1' \equiv \frac{dy_1}{dx_1} = \frac{\frac{dy}{dx} + \tau \frac{d\eta}{dx}}{1 + \tau \frac{d\xi}{dx}} = \left(y' + \tau \frac{d\eta}{dx}\right)\left(1 - \tau \frac{d\xi}{dx}\right) + \cdots$$

or

$$y_1' = y' + \tau\left(\frac{d\eta}{dx} - y' \frac{d\xi}{dx}\right) = y' + \tau\eta'(x,y,y') \tag{1.13-12}$$

(1.13-12) with the definition (1.13-11) completes the infinitesimal transformation in (x,y,y'). We can thus write the infinitesimal generator U' which is the extension of U to the (x,y,y') space.

$$U'f \equiv \xi(x,y) \frac{\partial f}{\partial x} + \eta(x,y) \frac{\partial f}{\partial y} + \eta'(x,y,y') \frac{\partial f}{\partial (y')} \tag{1.13-13}$$

where (from (1.13-12))

$$\eta' \equiv \frac{d}{dx}\left[\eta(x,y)\right] - y' \frac{d}{dx}\left[\xi(x,y)\right] \tag{1.13-14}$$

or written out using (1.13-11)

$$\eta' \equiv \eta_x + y'\left(\eta_y - \xi_x\right) - \xi_y y'^2 . \tag{1.13-15}$$

In (1.13-15) the dependence on (x,y,y') is shown explicitly, while for actual calculations the shorthand notation of (1.13-14) may be useful.

<u>Example</u>: rotation group

$$Uf \equiv -y \frac{\partial f}{\partial x} + x \frac{\partial f}{\partial y}$$

$$\xi = -y, \quad \eta = x$$

$$\eta' = 1 + y'^2$$

$$U'f \equiv -y \frac{\partial f}{\partial x} + x \frac{\partial f}{\partial y} + (1+y'^2) \frac{\partial f}{\partial (y')}$$

check directly

$$x_1 = x \cos \tau - y \sin \tau \to x - y\delta\tau$$

$$y_1 = y \cos \tau + x \sin \tau \to y + x\delta\tau$$

$$y_1' = \frac{\sin \tau + y' \cos \tau}{\cos \tau - y' \sin \tau} \to \frac{\delta\tau + y'}{1 - y'\delta\tau} = (y'+\delta\tau)(1+y'\delta\tau) = y' + (1+y'^2)\delta\tau.$$

1.14. A Second Criterion That a First-Order Differential Equation Admits a Group

The new criterion is derived by considering the transformations in the (x,y,y') space and is a direct consequence of §1.12 and §1.13. A first-order equation can be represented by

$$\Omega(x,y,y') = 0. \tag{1.14-1}$$

In the plane (1.14-1) defines an (∞^1) of integral curves; through each (x,y) one integral curve passes with a slope y' given by (1.14-1), on each branch if necessary. However, in (x,y,y') space (1.14-1) defines a surface (generated by integral curves). A differential equation admits a given group if the integral curves are taken into one another by the members of the group and this idea can be expressed infinitesimally for the group. In (x,y,y') space, in order to admit a given group it is necessary that the surface expressed by (1.14-1) admit the extended transformation (1.13-13) of the given point transformation. The surface goes into itself. That is, according to (1.12-28), the invariance criterion for a surface, (1.14-1) admits a given group (ξ,η) if and only if (cf. 1.13-13, 14, 15)

$$U'\Omega \equiv \xi \frac{\partial \Omega}{\partial x} + \eta \frac{\partial \Omega}{\partial y} + \eta' \frac{\partial \Omega}{\partial (y')} = 0 \text{ when } \Omega = 0 \text{ for all } x,y \tag{1.14-2}$$

where

$$\eta' \equiv \eta_x + (\eta_y - \xi_x)y' - \xi_y y'^2.$$

We assume that not all $(\Omega_x, \Omega_y, \Omega_{y'})$ are zero in the representation (1.14-1).

<u>Exercise</u>: Work out the criterion for $\Omega \equiv y' - \omega(x,y) = 0$.

Answer: $\eta_x + (\eta_y - \xi_x)\omega - \xi_y \omega^2 = \xi\omega_x + \eta\omega_y$.

<u>Example</u>: Differential equations for the family of straight lines tangent to the unit circle - differential equations for this family certainly admit the group of rotations

$$Uf \equiv -y \frac{\partial f}{\partial x} + x \frac{\partial f}{\partial y} .$$

These straight lines have the equation $ax + by = 1$ with $a^2 + b^2 = 1$; the differential equation can be found by elimination of the parameters (a,b).

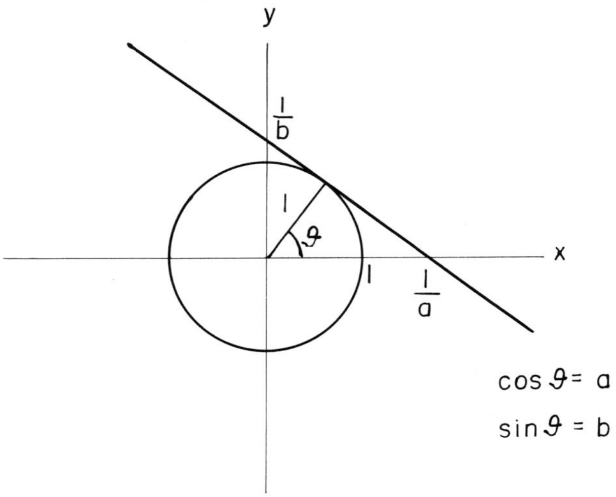

Figure 1.14-1

$$a + by' = 0$$

so

$$b = \frac{1}{y - xy'}, \qquad a = -\frac{y'}{y - xy'}$$

and the differential equation is

$$a^2 + b^2 = \frac{y'^2}{(y-xy')^2} + \frac{1}{(y-xy')^2} = 1$$

Hence

$$\Omega = y'^2 + 1 - (y-xy')^2 = 0$$

$$\Omega = 1 - y^2 + 2xyy' + (1-x^2)y'^2 = 0$$

check: invariance under extended transformation (see previous example; §1.13).

$$U'f \equiv -y\frac{\partial f}{\partial x} + x\frac{\partial f}{\partial y} + (1+y'^2)\frac{\partial f}{\partial y'}$$

invariance criterion

$$U'\Omega \equiv -y\frac{\partial \Omega}{\partial x} + x\frac{\partial \Omega}{\partial y} + (1+y'^2)\frac{\partial \Omega}{\partial (y')}$$

$$= -y[2yy'-2xy'^2] + x[-2y+2xy'] + (1+y'^2)[2xy + 2(1-x^2)y']$$

$$U'\Omega = +2y'[-y^2+xyy'+x^2+1-x^2+xyy'+(1-x^2)y'^2] = 2y'\Omega = 0 \text{ when } \Omega = 0.$$

Commutators: We may check that the use of the extended transformation yields the same criterion as before that a differential equation admits a given group, namely the commutator formula (1.9-17). Write the differential equation (1.14-1) in solved form:

$$\Omega \equiv Xy' - Y = 0, \quad ((X(x,y), Y(x,y)). \tag{1.14-3}$$

Then the criterion (1.14-2) becomes

$$\begin{aligned} U'\Omega \equiv\ & \xi\left\{\frac{\partial X}{\partial x}y' - \frac{\partial Y}{\partial x}\right\} + \eta\left\{\frac{\partial X}{\partial y}y' - \frac{\partial Y}{\partial y}\right\} \\ & + \left\{\frac{\partial \eta}{\partial x} + \left(\frac{\partial \eta}{\partial y} - \frac{\partial \xi}{\partial x}\right)y' - \frac{\partial \xi}{\partial y}y'^2\right\}X = 0 \quad \text{on} \quad \Omega = 0 \end{aligned} \tag{1.14-4}$$

Replacing y' by Y/X we have

$$\xi\left\{\frac{\partial X}{\partial x} Y - \frac{\partial Y}{\partial x} X\right\} + \eta\left\{\frac{\partial X}{\partial y} Y - \frac{\partial Y}{\partial y} X\right\} + \frac{\partial \eta}{\partial x} X^2 + \left(\frac{\partial \eta}{\partial y} - \frac{\partial \xi}{\partial x}\right) XY - \frac{\partial \xi}{\partial y} Y^2 = 0 \tag{1.14-5}$$

or using the operators U, A

$$U \equiv \xi \frac{\partial}{\partial x} + \eta \frac{\partial}{\partial y}, \qquad A \equiv X \frac{\partial}{\partial x} + Y \frac{\partial}{\partial y}$$

(1.14-5) becomes

$$YUX - XUY = YA\xi - XA\eta \tag{1.14-6}$$

Thus, directly

$$\frac{UX - A\xi}{X} = \frac{UY - A\eta}{Y} = \lambda(x,y), \text{ say.} \tag{1.14-7}$$

Note that (1.14-7) is exactly the criterion derived earlier (1.9-16) which leads to the commutator criterion (1.9-17).

1.15. Construction of All Differential Equations of First-Order Which Admit a Given Group

We are now in a position to construct the form of all differential equations of first-order which admit a given group and can thus be reduced to quadratures. We can show that when the path curves of the transformation are known the form of the equations depends only on a quadrature. This enables a dictionary approach to be carried out. A table of given groups and resulting differential equations can be constructed (see Table I).

However, note that the inverse problem of finding the group is not solved. A single equation (1.14-5) for (ξ, η) which must be satisfied for all (x,y), in the solved case (1.14-3), has already been derived. By working with this equation, which always has an infinite number of solutions, sometimes a group (ξ,η) can be discovered. But there is no systematic method for discovering a group, i.e., we are not able to find the general solution of (1.14-5) or systematically a particular solution of this equation.

Now, all differential equations

$$\Omega(x,y,y') = 0 \qquad (1.15\text{-}1)$$

which admit a given group, define a known surface in (x,y,y') space which is invariant under the extended group. We must find all surfaces which are swept out by path curves of the extended group U'. The basic result has already been derived (1.12-27 et seq.). If $u(x,y,y')$, $v(x,y,y')$ are two independent invariants then

$$W(u,v) = 0, \quad W \text{ arbitrary} \qquad (1.15\text{-}2)$$

is the general equation of an invariant surface. We may then write

$$v - w(u) = 0, \quad w \text{ arbitrary} \qquad (1.15\text{-}3)$$

(1.15-3) is thus the general form of differential equations admitting a given group. The result, however, can be simplified greatly. It is always possible to choose one of the invariants, say u, so that it does not depend on y' since ξ and η do not depend on y'. That is

$$u = u(x,y). \qquad (1.15\text{-}4)$$

Thus, for u to be an invariant

$$U'u = \xi \frac{\partial u}{\partial x} + \eta \frac{\partial u}{\partial y} = 0 \qquad (1.15\text{-}5)$$

(1.15-5) means that $u = \text{const.}$ on the path curves of the group

$$\left(\left(\frac{dy}{dx}\right) = \frac{\eta}{\xi}\right).$$

Now, if these path curves are known and represented by

$$u(x,y) = \text{const. on path curves} \tag{1.15-6}$$

then it is always possible to find $v(x,y,y')$ by quadrature. To demonstrate: the condition for invariance is the characteristic differential equations of (1.14-2) $U'\Omega = 0$.

$$\frac{dx}{\xi(x,y)} = \frac{dy}{\eta(x,y)} = \frac{d(y')}{\eta_x + (\eta_y - \xi_x)y' - \xi_y y'^2} \tag{1.15-7}$$

and $u(x,y) = \text{const.}$ is the integral of the first two of (1.15-7). Regarding this as known the second integral can be found by integration of

$$\frac{d(y')}{dx} = \frac{1}{\xi}\frac{\partial\eta}{\partial x} + \frac{1}{\xi}\left(\frac{\partial\eta}{\partial y} - \frac{\partial\xi}{\partial x}\right)y' - \frac{1}{\xi}\frac{\partial\xi}{\partial y}\,y'^2 \tag{1.15-8}$$

$$= F(x;\, c) + F_1(x;\, c)y' + F_2(x;\, c)y'^2 \tag{1.15-9}$$

where y has been eliminated by $u(x,y) = c$. Equation (1.15-9) is of Riccati type and, in general, cannot be solved by quadrature. In the special case that a particular solution is known the solution of the Riccati equation can be expressed by quadrature. This is the situation here. Note that the path curves and the associated direction field of the path curves admit the extended group and so give a particular solution. Let

$$y' = \frac{\eta}{\xi} = y'(x;\, c)\,. \tag{1.15-10}$$

Since

$$\xi y' - \eta = 0$$

$$\xi d(y') + \xi_x y'dx + \xi_y y'dy - \eta_x dx - \eta_y dy = 0$$

and this is nothing but the Riccati equation (1.15-8). Thus, writing

$$\frac{\eta}{\xi} = g(x;\ c) \tag{1.15-11}$$

$$\frac{dg}{dx} = F(x;\ c) + F_1(x;\ c)g + F_2(x;\ c)g^2. \tag{1.15-12}$$

Then according to the usual method for solving the Riccati equation (1.15-9) in general (from its equivalence to a second order linear differential equation), etc.

$$y' = g(x;\ c) + \frac{1}{h(x)}\ . \tag{1.15-13}$$

We have

$$\frac{d(y')}{dx} = \frac{dg}{dx} - \frac{1}{h^2}\frac{dh}{dx} = F(x;\ c) + F_1(x;\ c)\left[g + \frac{1}{h}\right] + F_2(x;\ c)\left[g + \frac{1}{h}\right]^2 .$$

Using (1.15-12) we obtain the following linear equation for $h(x)$:

$$\frac{dh}{dx} = -(F_1+2gF_2)h - F_2. \tag{1.15-14}$$

The linear equation can be solved by quadrature;

$$h = H(x;\ c;\ \gamma) \tag{1.15-15}$$

where γ is a constant of integration. $v(\gamma)$ is an arbitrary function. Thus (1.15-13) becomes

$$y' = \frac{\eta(x,y)}{\xi(x,y)} + \frac{1}{H(x;\ c;\ \gamma)}\ , \quad v = v(\gamma) \tag{1.15-16}$$

or $v(x,y,y')$ is found.

<u>In summary</u>: The general form of first-order differential equations admitting a given group (ξ,η) is

$$v(x,y,y') = w\left[u(x,y)\right], \quad w \text{ arbitrary} \tag{1.15-17}$$

where $u(x,y)$, $v(x,y,y')$ are independent integrals of $Uu, U'v$.

That is,

$$Uu = \xi \frac{\partial u}{\partial x} + \eta \frac{\partial u}{\partial y} = 0 \quad \text{for all} \quad x,y \tag{1.15-18}$$

$$U'v = \xi \frac{\partial v}{\partial x} + \eta \frac{\partial v}{\partial y} + \eta' \frac{\partial y}{\partial (y')} = 0 \quad \text{for all} \quad x,y,y' \tag{1.15-19}$$

with

$$\eta' \equiv \eta_x + (\eta_y - \xi_x)y' - \xi_y y'^2$$

or equivalently, u = const., v = const. are the integrals of the characteristic system:

$$\frac{dx}{\xi(x,y)} = \frac{dy}{\eta(x,y)} = \frac{d(y')}{\eta_x + (\eta_y - \xi_x)y' - \xi_y y'^2} . \tag{1.15-20}$$

Once $u(x,y)$ is known explicitly it is always possible to find $v(x,y,y')$ by quadrature, as in (1.15-16) above. Sometimes it is easier to calculate v directly. An alternative approach leading to the same result is the use of canonical coordinates.

<u>Canonical coordinates</u>: If the path curves

$$u(x,y) = c \tag{1.15-21}$$

are known then canonical coordinates

$$\begin{aligned} r &= r(x,y) \\ s &= s(x,y) \end{aligned} \tag{1.15-22}$$

can be introduced by quadrature (cf. 1.5-19, 1.5-20) and in these coordinates the group is the translation group

$$Uf \equiv \frac{\partial f}{\partial s} . \tag{1.15-23}$$

All differential equations which admit the translation group can be written in the form

$$\frac{ds}{dr} - F(r) = 0. \tag{1.15-24}$$

That is, $F(r,s)$ cannot contain s, and (1.15-24) is evidently invariant under translation with respect to s. We can choose

$$r \equiv u(x,y) \tag{1.15-25}$$

and, in (1.15-24)

$$\frac{ds}{dx} = s_x + s_y y', \quad \frac{dr}{dx} = r_x + r_y y'$$

so

$$v(x,y,y') \equiv \frac{ds}{dr} = \frac{s_x + s_y y'}{r_x + r_y y'}. \tag{1.15-26}$$

Thus, (1.15-24) is of the form (1.15-17).

Examples:

(i) uniform stretching

$$Uf \equiv x\frac{\partial f}{\partial x} + y\frac{\partial f}{\partial y}$$

$$U'f \equiv x\frac{\partial f}{\partial x} + y\frac{\partial f}{\partial y}$$

$$\frac{dx}{x} = \frac{dy}{y} = \frac{d(y')}{0}$$

$$u = \frac{y}{x}, \quad v = y';$$

$$y' = w\left(\frac{y}{x}\right) \text{ is the general form}$$

(ii) rotation

$$Uf \equiv -y\frac{\partial f}{\partial x} + x\frac{\partial f}{\partial y}$$

$$U'f \equiv -y\frac{\partial f}{\partial x} + x\frac{\partial f}{\partial y} + (1+y'^2)\frac{\partial f}{\partial (y')}$$

$$\frac{dx}{-y} = \frac{dy}{x} = \frac{d(y')}{1 + (y')^2}$$

$$u(x,y) = \sqrt{x^2+y^2}$$

$$\frac{dy}{\sqrt{u^2-y^2}} = \frac{d(y')}{1+(y')^2}$$

or

$$\tan^{-1}y' = \sin^{-1}\frac{y}{u} + \tan^{-1}v\,.$$

$$v = \tan\left(\tan^{-1}y' - \tan^{-1}\frac{y}{x}\right) = \frac{y' - y/x}{1 + y'y/x} = \frac{xy' - y}{x + yy'}\,.$$

In general form

$$\frac{xy' - y}{x + yy'} = w\left(\sqrt{x^2+y^2}\right).$$

In canonical coordinates

$$r = \sqrt{x^2+y^2} \qquad r_x = \frac{x}{r}\,, \quad r_y = \frac{y}{r}$$

$$s = \theta = \tan^{-1}\frac{y}{x} \qquad \theta_x = \frac{-y}{r^2}\,, \quad \theta_y = \frac{x}{r^2}$$

$$\frac{ds}{dr} = \frac{\theta_x + \theta_y y'}{r_x + r_y y'} = \frac{-y + xy'}{r(x+yy')} = F(r)$$

or

$$\frac{xy' - y}{x + yy'} = rF(r) = w\left(\sqrt{x^2+y^2}\right).$$

TABLE I

Sample Table of Groups and Differential Equations

Group $Uf \equiv \xi \frac{\partial f}{\partial x} + \eta \frac{\partial f}{\partial y}$	Equation	Canonical Coordinates
$x \frac{\partial f}{\partial x} + y \frac{\partial f}{\partial y}$	$y' = F\left(\frac{y}{x}\right)$	$r = \frac{y}{x}, \quad s = \log x$
$x \frac{\partial f}{\partial x} - y \frac{\partial f}{\partial y}$	$y' = \frac{y}{x} F(xy)$	$r = xy, \quad s = \log x$
$x \frac{\partial f}{\partial x} + \alpha y \frac{\partial f}{\partial y}$	$y' = \frac{y}{x} F\left(\frac{y}{x^\alpha}\right)$	$r = \frac{y}{x^\alpha}, \quad s = \log x$
$\frac{x}{\phi(x)} \left(x \frac{\partial f}{\partial x} + y \frac{\partial f}{\partial y}\right)$	$y' = \frac{y}{x} + \frac{\phi}{x} F\left(\frac{y}{x}\right)$	$r = \frac{y}{x}, \quad s = \int \frac{\phi}{x^2}\, dx$
$\frac{y}{\psi(y)} \left(x \frac{\partial f}{\partial x} + y \frac{\partial f}{\partial y}\right)$	$y' = \frac{y}{x} + \frac{y'}{x} \psi(y) F\left(\frac{y}{x}\right)$	$r = \frac{y}{x}, \quad s = \int \frac{\psi}{y^2}\, dy$
$\frac{1}{x\phi(x)} \left(x \frac{\partial f}{\partial x} - y \frac{\partial f}{\partial y}\right)$	$y' = -\frac{y}{x} + \frac{\phi(x)}{x} F(xy)$	$r = xy, \quad s = \int \phi\, dx$
$x^{n-k} \left(x \frac{\partial f}{\partial x} + ny \frac{\partial f}{\partial y}\right)$	$y' = \frac{ny}{x} + x^{k-1} F\left(\frac{y}{x^n}\right)$	$r = \frac{y}{x^n}, \quad s = x^{k-n}, \; k \neq n$
$x^{-a} \left(x \frac{\partial f}{\partial x} + ay \frac{\partial f}{\partial y}\right)$	$y' - \frac{a}{x} y + \frac{b}{x} y^2 = cx^{2a-1}$ (Riccati Special Case)	$r = \frac{y}{x^a}, \quad s = x^a$

TABLE I (CONTINUED)

$\left[e^{-\int \phi(x)\,dx}\right]\frac{\partial f}{\partial y}$	$y' = -\phi(x)y + F(x)$	$r = x, \quad s = e^{\int \phi dx}y$
$-y\,\frac{\partial f}{\partial x} + x\,\frac{\partial f}{\partial y}$	$\frac{y - xy'}{x + yy'} = F\left(\sqrt{x^2+y^2}\right)$ or $x + yy' = \sqrt{1+y'^2}\,F\left(\sqrt{x^2+y^2}\right)$	$r = \sqrt{x^2+y^2}, \quad s = \theta = \tan^{-1}\frac{y}{x}$
$\left[y^{\alpha}e^{(\alpha-1)\int \phi(x)\,dx}\right]\frac{\partial f}{\partial y}$	$y' = -\phi(x)y + F(x)y^{\alpha}$	$r = x, \quad s = e^{-(\alpha-1)\int \phi dx}y^{1-\alpha}$

1.16. Criterion That a Second-Order Differential Equation Admits a Group

The criterion that a second-order differential equation is invariant under a given one-parameter group of transformations is a simple extension of the ideas of the preceding two sections. Use it made of the twice-extended infinitesimal transformation of the group and the problem is studied in the (x,y,y',y'') space.

Let us first note that a second-order differential equation is equivalent to a two-parameter family (∞^2) of curves in the plane; and we are interested in such families of curves which are invariant under a group of transformations. To study this, represent the curves by

$$\omega(x,y;\ \alpha,\beta) = 0 \tag{1.16-1}$$

where (α,β) are essential parameters and (1.16-1) is a two-parameter family of curves in the plane. Under the point transformation

$$\begin{aligned} x_1 &= \phi(x,y) \\ y_1 &= \psi(x,y) \end{aligned} \tag{1.16-2}$$

curves go into curves. For invariance, curves of the family must go into curves of the family. That is, it must be possible to find α_1, β_1, depending only on α,β such that

$$\omega(x_1,y_1;\ \alpha_1,\beta_1) = 0,\ (\alpha_1(\alpha,\beta),\ \beta_1(\alpha,\beta)). \tag{1.16-3}$$

<u>Example</u>: Two-parameter family of straight lines in the plane go into straight lines under rotation.

$$\omega(x,y;\ \alpha,\beta) = y - \alpha x - \beta = 0$$

$$x = x_1 \cos\theta + y_1 \sin\theta$$

$$y = y_1 \cos\theta - x_1 \sin\theta$$

$$y - \alpha x - \beta = (y_1 \cos\theta - x_1 \sin\theta) - \alpha(x_1 \cos\theta + y_1 \sin\theta) - \beta$$
$$= y_1[\cos\theta - \alpha \sin\theta] - x_1[\sin\theta + \alpha\cos\theta] - \beta$$

$$y - \alpha x - \beta = 0 \quad \text{implies} \quad y_1 - \alpha_1 x_1 - \beta_1 = 0$$

where

$$\alpha_1 = \frac{\sin\theta + \alpha\cos\theta}{\cos\theta - \alpha\sin\theta}, \quad \beta_1 = \frac{\beta}{\cos\theta - \alpha\sin\theta}. \tag{1.16-4}$$

Now, the differential equation of the two-parameter family of curves can be obtained for (1.16-1) by differentiation and elimination of the parameters. Differentiating along a curve $y(x)$, we have

$$\frac{d\omega(x,y)}{dx} = \frac{\partial\omega}{\partial x} + \frac{\partial\omega}{\partial y} y'. \tag{1.16-5}$$

$$\frac{d^2\omega}{dx^2} = \frac{\partial^2\omega}{\partial x^2} + 2\frac{\partial^2\omega}{\partial x\partial y} y' + \frac{\partial^2\omega}{\partial y^2} y'^2 + \frac{\partial\omega}{\partial y} y''. \tag{1.16-6}$$

Elimination of (α,β) from (1.16-1, 5, 6) yields the differential equation in the form

$$\Omega(x,y,y',y'') = 0. \tag{1.16-7}$$

But, for invariance (cf. 1.16-3), the form of the differential equation of the two-parameter family of curves must read the same in the new variables, that is

$$\Omega(x_1,y_1,y_1',y_1'') = 0 \quad \text{for invariance.} \tag{1.16-8}$$

Then, families of integral curves go into families of integral curves.

The formal rules for the extension of the point transformation (1.16-2) are easily worked out by considering not only how points and associated tangent vectors (to given curves) transform but also how the local curvature transforms. The method is the same as was shown in detail in §1.13. Note that if

$$\begin{aligned} x_1 &= \phi(x,y) \\ y_1 &= \psi(x,y) \qquad\qquad (1.16\text{-}9) \\ y_1' &= \frac{\psi_x + \psi_y y'}{\phi_x + \phi_y y'} = \chi(x,y,y'),\ \text{say} \end{aligned}$$

then the twice-extended transformation is given by

$$y_1'' = \frac{d}{dx_1}\left(\frac{dy_1}{dx_1}\right) = \frac{\chi_x + \chi_y y' + \chi_{(y')} y''}{\phi_x + \phi_y y'} .$$

We note that any coordinates may be used and that if a two-parameter family of curves and/or its corresponding differential equation admits a transformation in one system of coordinates it will admit the same transformation when both D.E. and transformation are expressed in new coordinates. It is evident that if (1.16-2) depends on a parameter and represents a one-parameter group of transformations then the twice-extended transformation also has the group property.

Extended Infinitesimal Transformation.

In §1.13, the once-extended infinitesimal transformation was constructed and now by induction the twice-extended transformation can be written down. We have

$$Uf \equiv \xi(x,y)\frac{\partial f}{\partial x} + \eta(x,y)\frac{\partial f}{\partial y}\ ; \quad \text{for } f(x,y) \qquad (1.16\text{-}10)$$

$$U'f \equiv \xi(x,y)\frac{\partial f}{\partial x} + \eta(x,y)\frac{\partial f}{\partial y} + \eta'(x,y,y')\frac{\partial f}{\partial(y')}\ ; \ \text{for } f(x,y,y') \qquad (1.16\text{-}11)$$

where

$$\eta' \equiv \frac{d\eta(x,y)}{dx} - y'\frac{d\xi}{dx} \qquad (1.16\text{-}12)$$

the derivative being taken along the curve $y(x)$ whose slope is y'. In the same way

$$U''f \equiv \xi(x,y)\frac{\partial f}{\partial x} + \eta(x,y)\frac{\partial f}{\partial y} + \eta'(x,y,y')\frac{\partial f}{\partial(y')} \tag{1.16-13}$$
$$+ \eta''(x,y,y',y'')\frac{\partial f}{\partial(y'')} \; ; \text{ for } f(x,y,y',y'')$$

where now

$$\eta'' \equiv \frac{d\eta'(x,y,y')}{dx} - y''\frac{d\xi(x,y)}{dx} \tag{1.16-14}$$

with d/dx denoting the derivative along a curve whose slope is y' and curvature y''. The derivation of (1.16-14) follows from (1.13-10 ff)

$$y_1'' = \frac{dy_1'}{dx_1} = \frac{dy' + \tau d\eta'}{dx + \tau d\xi} = \frac{y'' + \tau\dfrac{d\eta'}{dx}}{1 + \tau\dfrac{d\xi}{dx}} = y'' + \tau\left(\frac{d\eta'}{dx} - y''\frac{d\xi}{dx}\right).$$

It is useful to write out the full formulas for η', η'' based on $\xi(x,y)$, $\eta(x,y)$

$$\eta' = \eta_x + (\eta_y - \xi_x)y' - \xi_y y'^2. \tag{1.16-15}$$

From (1.16-14)

$$\eta'' = \left[\eta_{xx} + \eta_{xy}y' + (\eta_{xy} - \xi_{xx})y' + (\eta_{yy} - \xi_{xy})y'^2 + (\eta_y - \xi_x)y''\right.$$
$$\left. - \xi_{xy}y'^2 - \xi_{yy}y'^3 - 2\xi_y y'y''\right] - y''\left[\xi_x + \xi_y y'\right]$$

or

$$\eta'' \equiv \eta_{xx} + (2\eta_{xy} - \xi_{xx})y' + (\eta_{yy} - 2\xi_{xy})y'^2 - \xi_{yy}y'^3 \tag{1.16-16}$$
$$+ (\eta_y - 2\xi_x)y'' - 3\xi_y y'y'' .$$

Note that η'' is linear in y''. The method of extension to higher orders is clear.

Now we can write down the criterion that a second-order differential equation is invariant under a given group. The hypersurface in $(x,y,y',y")$ space defined by (1.16-7) must transform into itself (cf. 1.16-8). Therefore, $U"\Omega = 0$ on this surface.

Criterion: A second-order differential equation

$$\Omega(x,y,y',y") = 0 \tag{1.16-17}$$

admits all the transformations of a one-parameter group

$$Uf \equiv \xi(x,y) \frac{\partial f}{\partial x} + \eta(x,y) \frac{\partial f}{\partial y}$$

when

$$U"\Omega \equiv \xi \frac{\partial \Omega}{\partial x} + \eta \frac{\partial \Omega}{\partial y} + \eta' \frac{\partial \Omega}{\partial (y')} + \eta" \frac{\partial \Omega}{\partial (y")} = 0$$

with the use of $\Omega = 0$ for all x,y,y'. When the differential equation is in solved form the criterion can be worked out more explicitly. Let

$$\Omega(x,y,y',y") \equiv y" - \omega(x,y,y') = 0. \tag{1.16-18}$$

Then $U"\Omega = 0$ becomes (cf. 1.16-15, 16)

$$\begin{aligned} -\xi \frac{\partial \omega}{\partial x} - \eta \frac{\partial \omega}{\partial y} - \left\{ \eta_x + (\eta_y - \xi_x) y' - \xi_y y'^2 \right\} \frac{\partial \omega}{\partial (y')} \\ + \eta_{xx} + (2\eta_{xy} - \xi_{xx}) y' + (\eta_{yy} - 2\xi_{xy}) y'^2 - \xi_{yy} y'^3 \\ + \left\{ \eta_y - 2\xi_x - 3\xi_y y' \right\} \omega(x,y,y') = 0 \quad \text{for all} \quad x,y,y' . \end{aligned} \tag{1.16-19}$$

When (1.16-19) holds the differential equation (1.16-18) admits the given group.

Example:

- straight lines in the plane
- differential equation $\Omega = y" = 0$
- should admit all projective transformations (cf. §1.7) especially (1.7-3). To check

-from (1.16-19)

$$\eta_{xx} + (2\eta_{xy}-\xi_{xx})y' + (\eta_{yy}-2\xi_{xy})y'^2 - \xi_{yy}y'^3 = 0$$

-thus we need, for invariance

$$\eta_{xx} = 0,\ 2\eta_{xy} - \xi_{xx} = 0,\ \eta_{yy} - 2\xi_{xy} = 0,\ \xi_{yy} = 0$$

-thus

$$\xi = a_0(x) + a_1(x)y,\quad \eta = b_0(y) + b_1(y)x$$

-and

$$b_0''(y) + xb_1''(y) = 2a_1'(x),\quad a_0''(x) + ya_1''(x) = 2b_1'(y)$$

-comparing these last two a_0'', b_0'' are const., a_1', b_1' are linear and the only solution is

$$a_0 = \alpha + \beta x + \kappa x^2,\ a_1 = \gamma + \lambda x,\ b_0 = \varepsilon + \theta y + \lambda y^2,$$
$$b_1 = \zeta + \kappa y$$

-or

$$\xi = \alpha + \beta x + \gamma y + \kappa x^2 + \lambda xy,\ \eta = \varepsilon + \zeta x + \theta y + \lambda y^2 + \kappa xy$$

-these are the projective transformations of the plane (1.7-2)

-there are eight independent infinitesimals $(\alpha,\beta,\gamma,\kappa,\lambda,\varepsilon,\zeta,\theta)$

$$\frac{\partial}{\partial x},\ x\frac{\partial}{\partial x},\ y\frac{\partial}{\partial x},\ x^2\frac{\partial}{\partial x} + xy\frac{\partial}{\partial y},\ xy\frac{\partial}{\partial x} + y^2\frac{\partial}{\partial y},\ \frac{\partial}{\partial y},\ x\frac{\partial}{\partial y},\ y\frac{\partial}{\partial y}$$

-linear combinations of these are possible, for example, rotation

$$-y\frac{\partial}{\partial x} + x\frac{\partial}{\partial y}\ .$$

Summary:

- we have thus shown that, in effect, non-trivially any second-order equation cannot admit more than eight independent one-parameter groups since $y'' = 0$ is the second-order differential equation with the richest symmetry. This is in contrast with first-order equations which admit an infinite number.
- there may be no group at all for a given second-order equation.

1.17. Construction of All Differential Equations of Second-Order Which Admit a Given Group

In a fashion parallel to that of §1.15 and using the material developed in §1.16, the form of second-order differential equations invariant under a given group is found in this section. The extension to higher order equations follows a similar path and is not presented. Use is made of the invariants derived earlier $\{u(x,y),\ v(x,y,y')\}$ and a new invariant $w(x,y,y',y'')$ is introduced. It is shown, as before, that if the path curves $u(x,y) = c$ of the group are known explicitly then the construction of further invariants, and hence of the general form of the invariant differential equation, is obtained by quadrature.

Let the given group define the once- and twice-extended operators

$$Uf \equiv \xi \frac{\partial f}{\partial x} + \eta \frac{\partial f}{\partial y} \tag{1.17-1}$$

$$U'f \equiv \xi \frac{\partial f}{\partial x} + \eta \frac{\partial f}{\partial y} + \eta' \frac{\partial f}{\partial (y')} \tag{1.17-2}$$

$$U''f \equiv \xi \frac{\partial f}{\partial x} + \eta \frac{\partial f}{\partial x} + \eta' \frac{\partial f}{\partial (y')} + \eta'' \frac{\partial f}{\partial (y'')} \tag{1.17-3}$$

where

$$\eta' \equiv \frac{d}{dx}\eta - y' \frac{d}{dx}\xi, \qquad \eta'' \equiv \frac{d}{dx}\eta' - y'' \frac{d}{dx}\xi$$

(cf. 1-16-15, 16).

Note that the partial differential equation in (x,y,y',y'')

$$U''f = 0 \tag{1.17-4}$$

has three independent integrals $(u,v,w,$ say) and that its general integral then is an arbitrary function F

$$F(u,v,w). \tag{1.17-5}$$

Further note that according to the general criterion of invariance (1.16-17) $U''\Omega = 0$ when $\Omega = 0$ for invariance. Thus the most general invariant differential equation is of the form

$$F(u,v,w) = 0. \tag{1.17-6}$$

Three independent invariants can be found from the characteristic differential equations of (1.17-4)

$$\frac{dx}{\xi(x,y)} = \frac{dy}{\eta(x,y)} = \frac{d(y')}{\eta'(x,y,y')} = \frac{d(y'')}{\eta''(x,y,y',y'')}. \tag{1.17-7}$$

If the path curves

$$u(x,y) = a \tag{1.17-8}$$

are known (and these are the integral of the first two of (1.17-7)) then a second invariant

$$v(x,y,y') = b \tag{1.17-9}$$

is found from the solution of the Riccati equation (1.15-8) (cf. 1.15-16). Then, eliminating y,y' in terms of a,b the last invariant $w(x,y,y',y'')$ can be found from

$$\frac{d(y'')}{dx} = \frac{\eta''(x,y(x;\ a),\ y'(x;\ a,b),\ y'')}{\xi(x,y(x;\ a))}. \tag{1.17-10}$$

Since η'' of (1.16-16) is linear in y'' the third invariant is found as the solution of a linear differential equation

$$w = W(y'',x;\ u,v) \tag{1.17-11}$$

and the general differential equation has the form (1.17-6).

However, another method can be used for the actual computation of the third invariant, and this method involves only differentiation. To see this, let

$v(x,y,y')$ be a differential invariant, that is $U'v = 0$

and

$u(x,y)$ be an invariant, that is $Uu = 0$.

Then, note that

$$v(x,y,y') - au(x,y) = b \tag{1.17-12}$$

is a first-order differential equation which is invariant under U, U', U''. If a is a fixed constant and we consider the family of differential equations for varying b, then we have a family of (∞^1) differential equations which are invariant under U. Each member of this family generates an (∞^1) family of integral curves. The totality of these curves is invariant under the given group. Considering both families (D.E. and integral curves) we have (∞^2) curves which are invariant and this must satisfy an invariant differential equation of the second-order. This differential equation can be found from (1.17-12) by differentiation processes and the elimination of parameters. From (1.17-12), d/dx along the curve $y(x)$, implies

$$\frac{\partial v}{\partial x} + \frac{\partial v}{\partial y} y' + \frac{\partial v}{\partial (y')} y'' - a\left(\frac{\partial u}{\partial x} + \frac{\partial u}{\partial y} y'\right) = 0. \tag{1.17-13}$$

This is invariant for any fixed constant a and we can write

$$a = \frac{\frac{\partial v}{\partial x} + \frac{\partial v}{\partial y} y' + \frac{\partial v}{\partial (y')} y''}{\frac{\partial u}{\partial x} + \frac{\partial u}{\partial y} y'} \tag{1.17-14}$$

or

$$W(y'',y',y,x) - a = 0.$$

Note that $U''(W-a) = 0$ or $U''W = 0$. The necessary third invariant is thus W, but we can write W as

$$w = W \equiv \frac{\frac{\partial v}{\partial x} + \frac{\partial v}{\partial y} y' + \frac{\partial v}{\partial (y')} y''}{\frac{\partial u}{\partial x} + \frac{\partial u}{\partial y} y'} \equiv \frac{dv}{du} . \tag{1.17-15}$$

The general form $F(u,v,w) = 0$ can thus be expressed as

$$\frac{dv}{du} = \Phi(u,v) . \tag{1.17-16}$$

<u>In summary</u>: The general second-order differential equation invariant under a given group can be expressed as

$$\frac{dv}{du} = \Phi(u,v), \qquad \Phi \text{ arbitrary}$$

where $u(x,y)$, $v(x,y,y')$ are the invariants (i.e., $Uu = 0$, $U'v = 0$) of

$$\frac{dx}{\xi} = \frac{dy}{\eta} = \frac{d(y')}{\eta_x + (\eta_y - \xi_x)y' - \xi_y y'^2}$$

and

$$\frac{dv}{du} = \frac{\frac{\partial v}{\partial x} + \frac{\partial v}{\partial y} y' + \frac{\partial v}{\partial (y')} y''}{\frac{\partial u}{\partial x} + \frac{\partial u}{\partial y} y'} .$$

The following remark is important. Not only do the above considerations give the general form of the second-order differential equation which admits a given group but also show exactly how the second-order differential equation is reduced to a first-order equation, namely (1.17-16). All that is necessary for this to be carried out explicitly is a knowledge of the path curves $u(x,y) = a$ and a quadrature to find v. From a study of the integral curves of

(1.17-16) complete qualitative information about the solutions can be found. Alternatively, (1.17-16) may admit further groups (this will be the case when the original second-order equation $\Omega(x,y,y',y'') = 0$ admits two independent groups, i.e., a two-parameter group) so that a complete reduction to quadrature is possible. Note that if the integrals of (1.17-16) are found

$$G(u(x,y),\ v(x,y,y')) = \text{const.} \tag{1.17-17}$$

then a first-order differential equation must be integrated to find the complete solution.

Examples:

(i) Perspective or Uniform Stretching Transformation

$$Uf \equiv x\frac{\partial f}{\partial x} + y\frac{\partial f}{\partial y},\quad \eta' = \eta_x + (\eta_y - \xi_x)y' - \xi_y y'^2 = 0$$

$$U'f \equiv x\frac{\partial f}{\partial x} + y\frac{\partial f}{\partial x}\ .$$

characteristic equations:

$$\frac{dx}{x} = \frac{dy}{y} = \frac{d(y')}{0}\ .$$

invariants:

$$u = \frac{y}{x},\quad v = y',\quad w \equiv \frac{dv}{du} = \frac{y''}{\frac{y'}{x} - \frac{y}{x^2}} = \frac{xy''}{v - u}\ .$$

invariant differential equation:

$$W(u,v,w) = 0 \quad \text{or} \quad F(xy'',y',\frac{y}{x}) = 0.$$

in solved form:

$$xy'' - \phi(\frac{y}{x},y') = 0 \qquad \text{(A)}$$

For integration of an equation of the form (A) we introduce invariant coordinates u,v determined by the perspective group $u = y/x$, $v = y'$

$$\frac{dv}{dx} = y'', \quad \frac{dv}{du} = \frac{xy''}{v - u} = \frac{\phi(u,v)}{v - u} \qquad \text{(B)}$$

then

$$\frac{du}{dx} = \frac{y'}{x} - \frac{y}{x^2} = \frac{v - u}{x}$$

and

$$\frac{du}{v - u} = \frac{dx}{x}, \text{ mapping to } x \text{ along an integral curve}$$

alternatively:

$v = F(u;\ c)$ is the integral of (B)

or

$y' = F(\frac{y}{x}\ ;\ c)$ (this also admits the group and hence has an integrating factor when F is known explicitly).

(ii) The Linear Equation

$$\Omega \equiv y'' + A(x)y' + B(x)y = 0$$

admits the affine group of stretching of y:

$$y_1 = \alpha y$$

$$x_1 = x$$

$$Uf \equiv y\,\frac{\partial f}{\partial y}, \quad \eta' = \eta_x + (\eta_y - \xi_x)y' - \xi_y y'^2 = y'$$

$$U'f \equiv y\,\frac{\partial f}{\partial y} + y'\,\frac{\partial f}{\partial (y')}\ .$$

characteristic equations:

$$\frac{dx}{(0)} = \frac{dy}{y} = \frac{d(y')}{(y')}$$

invariants:

$$u(x,y) = x, \quad v(x,y) = \frac{y'}{y}, \quad w = \frac{dv}{du} = \frac{\frac{y''}{y} - \frac{y'^2}{y^2}}{1}$$

invariant equation:

$$\frac{dv}{du} = \frac{y''}{y} - \frac{y'^2}{y^2} = \frac{-A(x)y'}{y} - B(x) - v^2$$

or

$$\frac{dv}{du} + v^2 = -A(u)v - B(u), \quad \text{Riccati equation}$$

Thus, every second-order linear equation is equivalent to a first-order non-linear Riccati equation.

(iii) Find the form of all second-order differential equations which admit the projective conformal group.

projective conformal group has (cf. 1.7-11):

$$\xi = \alpha + \beta x + \gamma y$$

$$\eta = \varepsilon - \gamma x + \beta y$$

$$U \equiv \xi \frac{\partial}{\partial x} + \eta \frac{\partial}{\partial y} \ .$$

calculations of invariants is simpler in polar coordinates:

$$r = \sqrt{(x-x_0)^2 + (y-y_0)^2}, \qquad \theta = \tan^{-1} \frac{y - y_0}{x - x_0}$$

where the origin is shifted so that

$$\xi = \beta(x-x_0) + \gamma(y-y_0), \quad \eta = -\gamma(x-x_0) + \beta(y-y_0)$$

in new coordinates (cf. 1.5-13):

$$Uf = Ur\,\frac{\partial f}{\partial r} + U\theta\,\frac{\partial f}{\partial \theta}$$

where

$$Ur = \left[\beta(x-x_0) + \gamma(y-y_0)\right]\frac{\partial r}{\partial x} + \left[-\gamma(x-x_0) + \beta(y-y_0)\right]\frac{\partial r}{\partial y} = \beta r$$

$$U\theta = \left[\beta(x-x_0) + \gamma(y-y_0)\right]\frac{\partial \theta}{\partial x} + \left[-\gamma(x-x_0) + \beta(y-y_0)\right]\frac{\partial \theta}{\partial y} = -\gamma$$

$$Uf \equiv \beta r\,\frac{\partial f}{\partial r} - \gamma\,\frac{\partial f}{\partial \theta}$$

$$U'f = \beta r\,\frac{\partial f}{\partial r} - \gamma\,\frac{\partial f}{\partial \theta} - \beta r'\,\frac{\partial f}{\partial (r')} \qquad \text{(cf. 1.13-15)}$$

where $r' = \dfrac{dr}{d\theta}$.

characteristic differential equations:

$$\frac{dr}{\beta r} = \frac{d\theta}{-\gamma} = \frac{d(r')}{-\beta r'}\,.$$

invariants:

$$u(r,\theta) = re^{\frac{\beta\theta}{\gamma}}$$

$$v(r',r,\theta) = r'e^{-\frac{\beta\theta}{\gamma}}\,.$$

general form of first-order invariant equations:

$$v = F(u); \quad r' = e^{\frac{\beta\theta}{\gamma}}\,F\left(re^{\frac{\beta\theta}{\gamma}}\right).$$

general form of second-order invariant equations:

$$\frac{dv}{du} = F(u,v)$$

$$\frac{r''e^{-\frac{\beta\theta}{\gamma}} - \beta/\gamma\; r'e^{-\frac{\beta\theta}{\gamma}}}{r'e^{\frac{\beta\theta}{\gamma}} + \frac{\beta}{\gamma}\,re^{\frac{\beta\theta}{\gamma}}} = F\left(re^{\frac{\beta\theta}{\gamma}}, r'e^{-\frac{\beta\theta}{\gamma}}\right).$$

other equivalent invariants:

$$u(r,\theta) = re^{\frac{\beta\theta}{\gamma}} \qquad v = rr'$$

$$rr' = F\left(re^{\frac{\beta\theta}{\gamma}}\right) \qquad \text{first-order}$$

$$\frac{dv}{du} = \frac{rr'' + r'^2}{r'e^{\frac{\beta\theta}{\gamma}} + \frac{\beta}{\gamma} re^{\frac{\beta\theta}{\gamma}}} = F\left(re^{\frac{\beta\theta}{\gamma}}, rr'\right)$$

or

$$rr'' + r'^2 = \left(\frac{r'}{r} + \frac{\beta}{\gamma}\right) G\left(rr', re^{\frac{\beta\theta}{\gamma}}\right); \quad \text{second-order.}$$

Canonical coordinates: Canonical coordinates can also be used to write the general form of second-order differential equations invariant under a given group.

Let $r(x,y)$, $s(x,y)$ be the canonical coordinates corresponding to U defined by

$$Ur = 0, \quad Us = 1. \tag{1.17-18}$$

Details are given in (1.5-17 ff). The group is one of translation in (r,s)

$$Uf \equiv \frac{\partial f}{\partial s}$$

$$U'f \equiv \frac{\partial f}{\partial s} .$$

The characteristic system of $(U'f = 0)$ is

$$\frac{dr}{0} = \frac{ds}{1} = \frac{ds'}{0}$$

so that the invariants are

$$u = r, \quad v = s' \equiv \frac{ds}{dr} . \tag{1.17-19}$$

The invariant first-order differential equation is thus:

$$\frac{ds}{dr} = \Phi(r) \tag{1.17-20}$$

and the invariant second-order equation is

$$\frac{dv}{du} \equiv \frac{d^2s}{dr^2} = F\left(r, \frac{ds}{dr}\right). \tag{1.17-21}$$

The equations (1.17-21) and (1.17-20) can be transformed back to (x,y) from the explicit knowledge of the canonical coordinates.

1.18. Examples of Application of the Method

In this section some typical examples, arising in different physical contexts, are worked out. A complete understanding of the differential equation is obtained by the consistent use of group theory. In these examples the groups are to be found, but the examples are sufficiently simple that it is fairly clear how to do this. Thus the emphasis is on how the use of transformation theory fits into the study of a general problem.

Example 1.18-1: The Differential Equation of a Problem in Variational Calculus.[1]

The physical problem concerns the drag due to friction and air pressure on the nose of a slender body in high speed flight. With the simplest (and not necessarily very realistic assumptions) of Newtonian impact pressure and laminar flow skin friction, the following formulas for pressure and shear stress are obtained:

$$\text{pressure} \qquad p - p_\infty = \rho_\infty U^2 \delta^2 F'^2(x) \tag{1.18-1}$$

$$\text{stress} \qquad \tau = \frac{k}{\sqrt{x}} \frac{p_\infty U^2}{2}. \tag{1.18-2}$$

---

[1] Theory of Optimum Aerodynamic Slopes, (A. Miele, Ed.), Chapter 15, Academic Press, 1965.

Here the cylindrical radius of the nose ogive is $r = \delta F(x)$, $\delta << 1$,

$$F(0) = 0, \quad F(1) = 1, \quad p = \text{pressure},$$

$$\rho_\infty = \text{ambient density}, \; U = \text{flight speed},$$

$$p_\infty = \text{ambient pressure}.$$

Thus the drag of the nose is (approximately, using $\delta << 1$)

$$D = 2\pi\rho_\infty U^2\delta^2 \int_0^1 \left\{F'^3(x) + \frac{K^3}{\sqrt{x}}\right\} F(x)\,dx \tag{1.18-3}$$

where K = a parameter of similitude = $(k/2\delta^2)^{1/3}$. The problem of finding the shape which minimizes the drag for constraints of given δ and length is the problem of finding the shape function $F(x)$ which minimizes I

$$\text{where} \qquad I = \int_0^1 \left\{y'^3(x) + \frac{K^3}{\sqrt{x}}\right\} y\,dx, \quad y = F(x). \tag{1.18-4}$$

The Euler-Lagrange equation of this problem is

$$\Omega(x,y,y',y'') = 3yy'y'' + y'^3 - \frac{1}{2}\frac{K^3}{\sqrt{x}} = 0 \tag{1.18-5}$$

and the boundary conditions are $y(0) = 0$, $y(1) = 1$ (say). The existence of the solution as well as qualitative features are seen after the ideas of invariance are used. The differential equation (1.18-5) is invariant under a scaling of the form

$$\frac{y^3}{x^3} \sim \frac{1}{\sqrt{x}},$$

or

$$\begin{aligned} y_1 &= \alpha^{5/6} y \\ x_1 &= \alpha x \end{aligned} \; . \tag{1.18-6}$$

Infinitesimally

$$Uf = x \frac{\partial f}{\partial x} + \frac{5}{6} y \frac{\partial f}{\partial y}$$

$$U'f = x \frac{\partial f}{\partial x} + \frac{5}{6} y \frac{\partial f}{\partial y} - \frac{1}{6} y' \frac{\partial f}{\partial (y')}$$

and the characteristic differential equations are

$$\frac{dx}{x} = \frac{dy}{\frac{5}{6} y} = \frac{dy'}{\left(- \frac{1}{6} y'\right)} .$$

Hence, $u = x^{-5/6}y$, $v = x^{1/6}y'$. These invariant coordinates can also be obtained directly from the global group, i.e., $y_1/x_1^{5/6} = y/x^{5/6} =$ $=$ const. are path curves. In order to eliminate the parameter K, let

$$\left.\begin{aligned} Ks &= x^{-5/6} y(x) \\ Kt &= x^{1/6} \frac{dy}{dx} \end{aligned}\right\} y(1) = 1, \quad s_f = \frac{1}{K} . \tag{1.18-7}$$

The differential forms of (1.18-7) are

$$\left.\begin{aligned} \frac{ds}{dx} &= \frac{1}{x} \left\{ t - \frac{5}{6} s \right\} \\ \frac{dt}{dx} &= \frac{1}{x} \left\{ \frac{x^{7/6}}{K} \frac{d^2y}{dx^2} + \frac{1}{6} t \right\} \end{aligned}\right\} . \tag{1.18-8}$$

so that the map from (t,s) trajectory to x coordinate is

$$\frac{dx}{x} = \frac{ds}{t - \frac{5}{6} s} . \tag{1.18-9}$$

In terms of the (s,t) coordinates, (1.18-5) reads

$$\frac{dt}{ds} = \frac{\frac{1}{2} + \frac{1}{2}st^2 - t^3}{3st\left(t - \frac{5}{6}s\right)} . \tag{1.18-10}$$

The reduction of the problem to the study of the trajectory of a first-order equation (1.18-10) followed by a quadrature (1.18-9) has thus been carried out. A brief discussion follows. Only the quadrant $s > 0,\ t > 0$ need be considered. In these new coordinates, every nonsingular solution is characterized by the fact that the initial point is located at the origin of the st-plane or at infinity. Furthermore, the abscissa of the final point is given by

$$s_f = \frac{1}{K}$$

so that its location depends on the parameter K. The relationship (1.18-9) exhibits a singularity along the straight line

$$t = \frac{5}{6}s \tag{1.18-11}$$

which, therefore, separates the st-domain into the two regions indicated in Fig. 1.18-1. In the region above this line x increases as s increases; conversely, the region below is characterized by decreasing values of x.

The general form of the paths which are solutions of this differential equation is indicated in Fig. 1.18-1, where the arrows indicate the direction of increasing x. Notice that the isoclines of infinite slope are represented by the relationships

$$s = 0, \quad t = 0, \quad t = \frac{5}{6}s \tag{1.18-12}$$

while the isocline of zero slope is given by

$$s = 2\left(t - \frac{1}{2t^2}\right). \tag{1.18-13}$$

Also, the intersection of these isoclines in the ts-plane yields the singular points c and 0 whose coordinates are

$$c: \quad s_c = 1.293 \qquad 0: \quad s_0 = 0$$
$$t_c = 1.077 \qquad t_0 = 0.794$$

Critical solutions. The first particular case occurs when $s_f = s_c$. In this case, one can readily prove that $K = K_c$; that is, the friction parameter is equal to a critical value. The singular point c represents the entire solution in the st-plane, and the corresponding equation for the shape is the power law,

$$y(x) = x^{5/6}. \tag{1.18-14}$$

Subcritical solutions. The second particular case occurs when $s_f > s_c$; that is, when the friction coefficient satisfies the inequality $K < K_c$. Now, if the initial point were at the origin of the st-plane, the associated path would be located in the region limited by the isoclines of infinite slope and zero slope. Since this path cannot overshoot the singular point c it is not possible to satisfy the boundary conditions of the problem at the final point. This leaves only one alternative; the initial point is located at infinity in the st-plane, and the regular shape is represented by a path running from infinity to $s = s_f$.

In order to investigate the behavior of the solution for large values of s and t, equation (1.18-10) is now approximated as

$$\frac{dt}{ds} \sim \frac{t\left(\frac{1}{2}s - t\right)}{3s\left(t - \frac{5}{6}s\right)} \tag{1.18-15}$$

which admits a solution of the type

$$t = \mu s \tag{1.18-16}$$

provided that $\mu = 3/4$. Along the special path $t = 3s/4$, Equation (1.18-9) can be rewritten in the form

$$\frac{ds}{s} = -\frac{1}{12}\frac{dx}{x} \tag{1.18-17}$$

whose general solution is

$$s = C_2 x^{-1/12} \tag{1.18-18}$$

where C_2 is an integration constant.

As the variable s becomes infinitely large, the variable x tends to zero. Consequently, the point at infinity of the st-plane corresponds to the origin $x = 0$, $y = 0$. Furthermore, one deduces that the relationship

$$y(x) = Cx^{3/4} \tag{1.18-19}$$

(where C is a constant) holds in the neighborhood of the origin of the ogive. Thus, the exponent of the shape of the body at the nose is independent of the friction parameter and is actually equal to that of the inviscid flow optimum shape.

As the friction parameter decreases, the final coordinate s_f increases and (1.18-19) holds over a larger and larger part of the body. In the limit, when $K \to 0$, the final coordinate s_f becomes infinitely large; consequently, equation (1.18-19) holds over the entire body.

Supercritical solutions. The third and final particular case occurs when $s_f < s_c$; that is, when $K > K_c$. By means of a reasoning complementary to that of the previous section, one can exclude the possibility that the initial point be located at infinity in the st-plane. This leaves one alternative: the initial point is located at the origin of the st-plane. Since the isocline of infinite slope passes through the origin, there exists no path which issues from the origin of the st-plane and reaches the specified final point. Consequently, if the friction parameter exceeds the critical value, there exists no regular shape solution joining the specified end-points. However, regular shape solutions do exist which connect the singular point with the final point f. In order to investigate this question, the immediate neighborhood of the point is considered, and the differential equation (1.18-10) is approximated in the form

$$\frac{d(t-t_0)}{ds} \simeq \frac{s - 6(t-t_0)}{6s} . \qquad (1.18\text{-}20)$$

After considering that $s_0 = 0$, one deduces that this equation admits the particular solution

$$t - t_0 = \frac{s}{12} \qquad (1.18\text{-}21)$$

which implies that

$$\frac{dx}{x} = \frac{ds}{t_0 - \frac{3}{4}\, s} \qquad (1.18\text{-}22)$$

If the initial conditions $x = x_0$, $s = 0$ are accounted for, this differential equation admits the particular integral

$$\left(\frac{x}{x_0} \right)^{-3/4} = 1 - \frac{3}{4} \frac{s}{s_0} . \qquad (1.18\text{-}23)$$

Hence, if the friction parameter is supercritical, a regular shape solution exists in the interval $x_0 < x \le 1$. The remaining part of the body, corresponding to the interval $0 \le x \le x_0$, is a spike of zero thickness.

The analogous calculations can easily be carried out for a skin friction coefficient which varies as any power of distance from the nose.

The results of this example also suggest that a general study of the Euler-Lagrange equations as far as groups are concerned can be made.

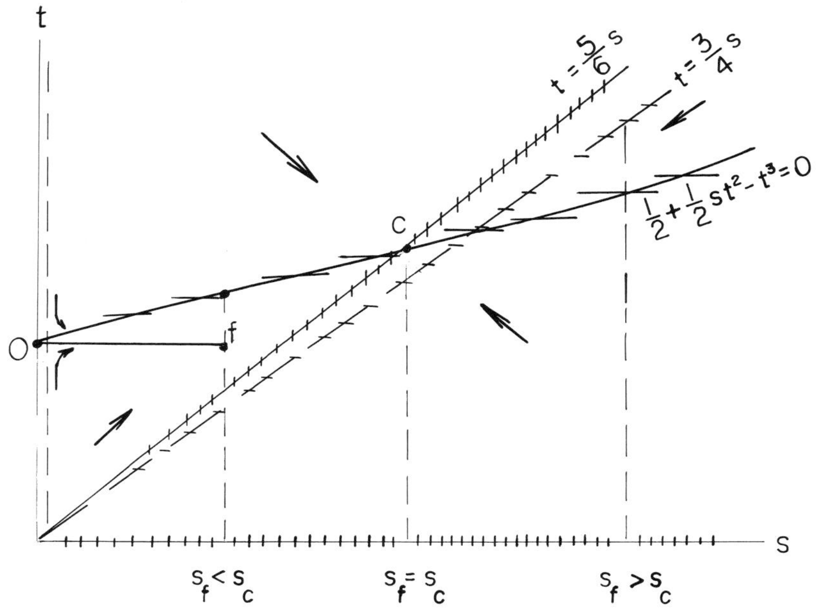

Figure 1.18-1

Example 1.18-2. Thomas-Fermi Equation

The Thomas-Fermi equation arises from a statistical model of a many electron atom. For a detailed discussion see, for example, Chapter 7 of "Intermediate Quantum Mechanics" by H. A. Bethe, Benjamin 1964. The use of transformation theory in this case allows the reduction to a first-order equation so that complete information about the structure of the solutions is obtained. In this way the existence of the solution under the given boundary conditions is demonstrated and a computation procedure can be outlined. These considerations simplify the mathematical discussion of Chapter 7, although of course no new results are obtained.

The Thomas-Fermi equation is the spherically symmetric version of the Poisson equation for the electric potential $1/e\ V_1$ outside the nucleus of a many electron atom:

$$\nabla^2 V_1 = -4\pi e^2 \rho. \tag{1.18-24}$$

Here $V_1 = V - \zeta$, $-V/e$ = potential, ζ = energy of the most energetic electrons, e = electronic charge, $-e\rho$ = charge density. The cloud of electrons is treated by Fermi-Dirac free particle statistics. From these statistical considerations it is shown that

$$\rho = \frac{1}{3\pi^2} \frac{(2m)^{3/2}}{\hbar^3} (-V_1)^{3/2} \tag{1.18-25}$$

where m = mass of atom, $2\pi\hbar$ = Planck's constant, so that (1.18-24) becomes a nonlinear ordinary differential equation:

$$\frac{1}{r} \frac{d^2}{dr^2} (rV_1) = -\frac{4e^2}{3\pi\hbar^3} (2m)^{3/2} (-V_1)^{3/2}. \tag{1.18-26}$$

The solution is sought for $r > 0$, but as $r \to 0$ we must have the potential of the concentrated source (nucleus) at the origin

$$V_1 \to -\frac{Ze^2}{r} \quad \text{as} \quad r \to 0, \quad Z = \text{atomic number}. \tag{1.18-27}$$

Suitable dimensionless variables can be introduced:

$$y = -\frac{rV_1}{Ze^2}, \qquad x = \frac{r}{b} \tag{1.18-28}$$

with

$$b = \frac{(3\pi)^{2/3}}{2^{7/3}} \frac{\hbar^2}{me^2} Z^{-1/3} = .885\ a_0 Z^{-1/3}.$$

The characteristic length is $a_0 = \hbar^2/me^2$ = first Bohr radius. For neutral free atoms a boundary condition at infinity is also defined. The "surface" of the atom corresponds to $r \to \infty$ where $\rho \to 0$. (Actually $\zeta = 0$). No net charge demands

$$rV_1 \to 0 \quad \text{as} \quad r \to \infty. \tag{1.18-29}$$

Thus, for neutral atoms the problem is

$$\frac{d^2y}{dx^2} = \frac{y^{3/2}}{\sqrt{x}} \qquad 0 < x < \infty \tag{1.18-30}$$

with

$$y(0) = 1 \tag{1.18-31}$$

$$y(\infty) = 0. \tag{1.18-32}$$

The equation (1.18-30) scales under stretching transformations as

$$\frac{y}{x^2} \sim \frac{y^{3/2}}{x^{1/2}}$$

or

$$x^3 y \sim \text{const.}$$

That is, the group is

$$y_1 = \alpha^3 y, \quad x_1 = \frac{1}{\alpha} x. \tag{1.18-33}$$

The corresponding invariant coordinates (cf. previous example) are

$$u = x^3 y, \quad v = x^4 y' = x^4 \frac{dy}{dx}. \tag{1.18-34}$$

Then

$$\frac{du}{dx} = x^3 y' + 3x^2 y = \frac{v + 3u}{x}$$

$$\frac{dv}{dx} = x^4 y'' + 4x^3 y' = \frac{x^5 y'' + 4v}{x} = \frac{x^5 \frac{y^{3/2}}{\sqrt{x}} + 4v}{x} = \frac{u^{3/2} + 4v}{x} .$$

The first-order equation to be studied is

$$\frac{dv}{du} = \frac{4v + u^{3/2}}{v + 3u} \tag{1.18-35}$$

and the mapping to x along an integral curve of (1.18-35) is given by

$$\frac{dx}{x} = \frac{du}{v + 3u} = \frac{dv}{4v + u^{3/2}} . \tag{1.18-36}$$

A sketch of the paths of (1.18-35) is given in Fig. (1.18-2). We are interested in $u > 0$, $v < 0$. The isoclines of zero slope $(4v + u^{3/2} = 0)$ and infinite slope $(v + 3u = 0)$ are drawn as well as some representative paths. The arrows on the paths indicate the direction in which x increases. There is one singular point P of interest where

$$4v_P + u_P^{3/2} = 0$$

$$v_P + 3u_P = 0$$

or $(u_P = 144,\ v_P = -432)$ and, as usual, the singular point represents one exceptional solution y_E of (1.18-30)

$$y_E = \frac{144}{x^3} . \tag{1.18-37}$$

The behavior near the origin can be obtained from the local form of (1.18-35). Many paths run into the origin between the isoclines, and (it turns out) that on these paths $u >> v >> u^{3/2}$ so that (1.18-35) is approximated by

$$\frac{dv}{du} = \frac{4}{3}\frac{v}{u} + \cdots . \tag{1.18-38}$$

Thus, near the origin, on all these paths

$$v = c_o u^{4/3} + \cdots , \tag{1.18-39}$$

c_o to be determined.

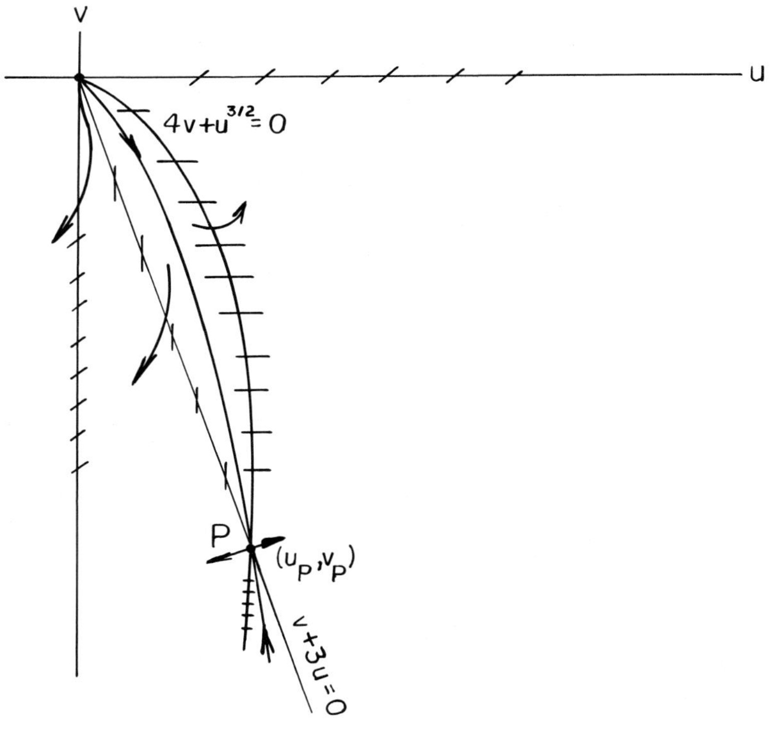

Figure 1.18-2

From the integration of the mapping formula (1.18-36)

$$\frac{dx}{x} = \frac{du}{3u}$$

we have

$$u = a_0 x^3 + \cdots = yx^3. \tag{1.18-40}$$

Thus, the origin of (u,v) corresponds to $x = 0$ and the boundary condition (1.18-31) determines the constant of integration in the mapping back to x, namely

$$a_0 = 1. \tag{1.18-41}$$

The only path which has a chance to satisfy the boundary condition at infinity is the exceptional path running from the origin to the saddle point at P. The behavior as the solution approaches the saddle point along this path should approach that of the singular solution (1.18-37) and so satisfy the boundary condition at infinity. To verify this we can study the neighborhood of the singular point P. Let

$$\begin{aligned} u &= u_P + u^* \\ v &= v_P + v^* \end{aligned} \tag{1.18-42}$$

so (1.18-35) becomes

$$\begin{aligned} \frac{dv^*}{du^*} &= \frac{4(v_P+v^*) + (u_P+u^*)^{3/2}}{v_P + v^* + 3u_P + 3u^*} \\ &= \frac{4v^* + \frac{3}{2}u_P^{1/2}u^*}{v^* + 3u^*}, \quad u_P^{1/2} = 12. \end{aligned} \tag{1.18-43}$$

According to the usual tests and our qualitative considerations the singular point is a saddle point and the exceptional paths can be found by letting

$$v^* = \kappa u^*. \tag{1.18-44}$$

Then

$$\kappa = \frac{4\kappa + 18}{\kappa + 3}$$

or

$$\kappa^2 - \kappa - 18 = 0. \tag{1.18-45}$$

The roots are

$$\kappa_{1,2} = \frac{1}{2} \pm \sqrt{\frac{1}{4} + 18}. \qquad (1.18\text{-}46)$$

The exceptional paths have positive and negative slopes respectively and lie in the quadrants defined by the isoclines. Let $-\kappa_2 = +\lambda = 3.76$. Thus, along the exceptional path running from the origin to the singular point P

$$v^* = -\lambda u^* \qquad (1.18\text{-}47)$$

and the mapping formula (1.18-36) shows that

$$\frac{dx}{x} = -\frac{du^*}{(\lambda-3)u^*}. \qquad (1.18\text{-}48)$$

Integration leads to

$$x = a_\infty (u^*)^{1/3-\lambda} \qquad (1.18\text{-}49)$$

so that, in fact, $x \to \infty$ as $u^* \to 0$. The constant (a_∞) in the mapping formula is not arbitrary but has already been found, in theory, from the considerations near the origin. The form of corrections to (1.18-37) is found from (1.18-49)

$$y \to \frac{1}{x^3}\left\{ 144 + \left(\frac{a_\infty}{x}\right)^{\lambda-3} + \cdots \right\}. \qquad (1.18\text{-}50)$$

For numerical calculations the following procedure defines the integration of the problem:

(1) Starting at the saddle point (u_P, v_P) integrate (1.18-35) along the exceptional path toward the origin, using (1.18-47) to get started. As a result the constant c_o in equation (1.18-39) is determined. The trajectory $v(u)$ is now established.

(2) Along this trajectory integrate the mapping formula (1.18-36)

$$\frac{dx}{x} = \frac{du}{v(u) + 3u}$$

using, for the constant of integration, a_0, as already determined.

The other paths in the (u,v)-plane which also represent solutions of the Thomas-Fermi equation can represent solutions for different conditions, e.g., ions of neutral atoms under pressure in which cases the solutions run only to a finite value x_0.

Example 1.18-3. Blasius Equation

This example shows the application of the same kind of reasoning to a higher-order equation. The ordinary differential equation arises from the similarity solution to a nonlinear partial differential equation and is of third-order. In this case, the third-order equation admits two independent invariances[2] so that by repeated application of the ideas developed earlier the solution of the problem is reduced to the study of the solutions of a first-order equation plus two quadratures. For the first-order equation it turns out, as is typical for so many cases, not all the paths need to be constructed, but only a certain exceptional path. In this special sense the problem is reduced entirely to quadratures.

The differential equation arises from that for a stream function in viscous incompressible flow past a semi-infinite flat plate. An asymptotic expansion of the solution of the Navier-Stokes equations is constructed in a similarity form. The method of arriving at this form is the subject of discussion in Part 2.

Here the stream function similarity form is:

$$\psi(x,y) = U\sqrt{\frac{x\nu}{U}}\, f(\eta), \quad \eta = y\sqrt{\frac{U}{\nu x}}$$

$$(x,y) = \text{Cartesian coordinates} \qquad (1.18\text{-}51)$$

$$U = \text{free stream velocity}, \quad \nu = \text{kinematic viscosity}.$$

---

[2] H. Weyl, On the Simplest Differential Equations of Boundary Layer Theory, Ann. of Math., 43, 2, pp. 381-407.

The velocity components are

$$u = \psi_y = Uf'(\eta)$$
$$v = -\psi_x = -\frac{1}{2}\sqrt{\frac{U\nu}{x}}\,\{f-\eta f'\} \tag{1.18-52}$$

and the skin-friction at the plate depends on

$$u_y = U\sqrt{\frac{U}{\nu x}}\,f''(\eta) \tag{1.18-53}$$

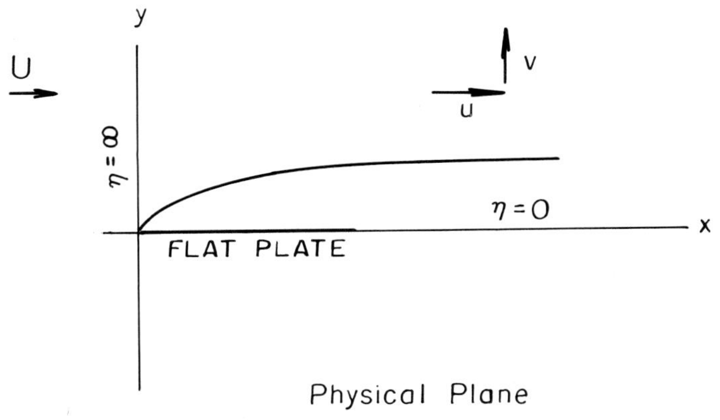

Figure 1.18-3

The momentum equation for the x-direction is

$$f''' + \frac{1}{2}ff'' = 0, \quad \text{Blasius equation.} \tag{1.18-54}$$

The boundary conditions are that the plate is a streamline and there is no slip:

$$f(0) = f'(0) = 0 \tag{1.18-55}$$

and the boundary condition of uniform flow at infinity (or at $x = 0$):

$$f(\eta) \to \eta \quad \text{as} \quad \eta \to \infty. \tag{1.18-56}$$

The equation (1.18-54) has two invariances, translation and scaling

$$\eta \to \eta + y_0 \quad \text{and} \quad f \sim \frac{1}{\eta}$$

which implies the possibility of reduction of the problem to the integration of a first-order equation followed by two quadratures.

A sketch of the expected course of the solution is given in Fig. 1.18-4.

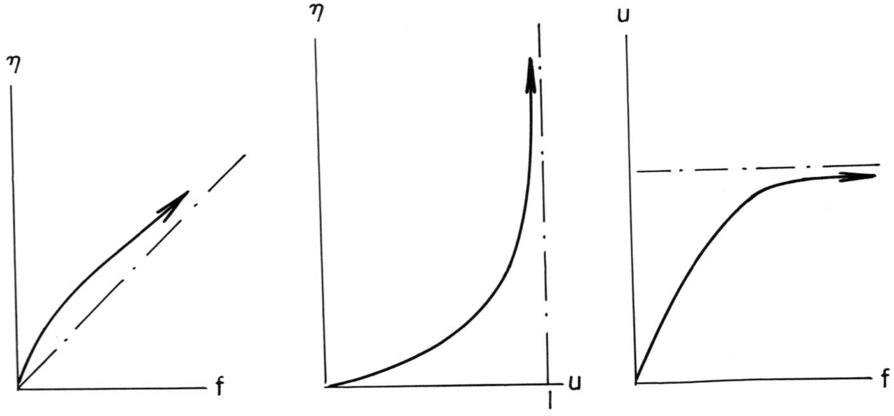

Figure 1.18-4

The translation invariance allows the reduction of order by the introduction of the derivative as a variable. This is the same idea as used in mechanics of autonomous systems where the invariance is

$(t \to t + t_0)$ and the velocity $dy/dt = v$, is used as a coordinate in addition to y (phase plane). Formally, the invariance here is

$$f_1 = f$$
$$\eta_1 = \eta + \alpha \tag{1.18-57}$$

The infinitesimal transformation is given by Ug and is equal to its extension, i.e.,

$$Ug = \frac{\partial g}{\partial \eta} = U'g.$$

The characteristic differential equations are

$$\frac{d\eta}{1} = \frac{df}{0} = \frac{d(f')}{0},$$

so that f, f' are invariant coordinates.

Let

$$w = \frac{df}{d\eta} \tag{1.18-58}$$

and study the equation in (w,f) coordinates. Thus

$$\frac{d^2f}{d\eta^2} = \frac{dw}{d\eta} = w\frac{dw}{df}, \quad \frac{d^2w}{d\eta^2} = w\frac{d}{df}\left\{w\frac{dw}{df}\right\}.$$

Hence, (1.18-54) becomes

$$w\frac{d}{df}\left\{w\frac{dw}{df}\right\} + \frac{1}{2}fw\frac{dw}{df} = 0$$

or

$$w\frac{d^2w}{df^2} + \left(\frac{dw}{df}\right)^2 + \frac{1}{2}f\frac{dw}{df} = 0. \tag{1.18-59}$$

The path on which $dw/df = \infty$ at $f = 0$ is desired so that $f''(0)$ is finite.

Equation (1.18-59) has a further scaling invariance

$$w \sim f^2 \tag{1.18-60}$$

corresponding to $f \sim \frac{1}{\eta}$ and the group is $(w_1 = \alpha^2 w,\ f_1 = \alpha f)$, so that suitable invariant coordinates are

$$s = w/f^2, \quad t = \frac{1}{f}\frac{dw}{df}\ . \tag{1.18-61}$$

The map from (s,t) to (w,f) is given differentially by

$$\frac{df}{f} = \frac{ds}{t - 2s} \tag{1.18-62}$$

and it is easily found that

$$\frac{d^2w}{df^2} = (t-2s)\frac{dt}{ds} + t. \tag{1.18-63}$$

In terms of the (s,t) coordinates, equation (1.18-59) is thus

$$\frac{dt}{ds} = \frac{t}{s}\left(\frac{\frac{1}{2} + t + s}{2s - t}\right). \tag{1.18-64}$$

The end points of $w(f)$ are given by

(i) $w \to 0,\quad f \to 0,\quad (s,t) \to \infty$

(ii) $w \to 1,\quad f \to \infty,\quad (s,t) \to 0.$

A qualitative picture of the paths is sketched in Figure 1.18-5.

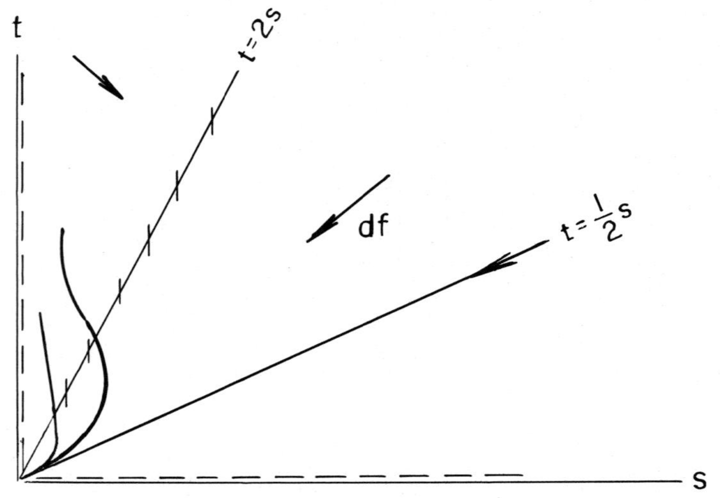

Figure 1.18-5

The arrows on the figure indicate the direction of increasing f.

Near infinity the exceptional path $t = \lambda s$, $\lambda = \text{const.}$ is possible if (cf. 1.18-64)

$$\lambda = \lambda\left(\frac{\lambda + 1}{2 - \lambda}\right) \quad \text{so that} \quad \lambda = \frac{1}{2}\,,\ t = \frac{1}{2}\,s. \tag{1.18-65}$$

On this exceptional path near infinity (1.18-62) shows

$$-\frac{2}{3}\frac{ds}{s} = \frac{df}{f} \tag{1.18-66}$$

so that

$$f = \frac{k_\infty}{s^{2/3}}\,,\quad k_\infty = \text{const.} \tag{1.18-67}$$

Thus

$$w = k_\infty^{3/2} f^{1/2} = \frac{df}{d\eta}$$

and

$$f = \frac{1}{4}\,k_\infty^3 \eta^2 + \cdots \quad \text{as} \quad \eta \to 0$$

in accord with the boundary conditions and the expectation $f''(0) = \text{const.} = \frac{1}{2}\,k_\infty^3$.

At the origin there is a higher order singularity, but the behavior of the paths can be approximated for $s >> t$ by (cf. 1.18-64)

$$\frac{dt}{ds} = \frac{1}{4}\frac{t}{s^2}\,. \tag{1.18-68}$$

Thus, the paths are approximately

$$t = k_0 e^{-1/4s}. \tag{1.18-69}$$

The mapping to f (1.18-62) is approximately

$$\frac{df}{f} = -\frac{1}{2}\frac{ds}{s} \tag{1.18-70}$$

so that

$$f(s) = \frac{d_0}{\sqrt{s}} \to \infty, \quad s \to 0. \tag{1.18-71}$$

Consideration of (1.18-61) shows that $d_0 = \sqrt{w}$ and that the boundary condition (1.18-56) is satisfied by the choice $d_0 = 1$. Thus

$$w = 1 = \frac{df}{d\eta}, \quad f \to \eta, \quad \eta \to \infty.$$

Theoretically, the construction of the solution has been accomplished by

(i) construction of the exceptional path, asymptotic to $t = \frac{1}{2} s$ at infinity,

(ii) evaluation of $f(s)$ along the path starting from $f = 1/\sqrt{s}$ as $s \to 0 \cdots$ leading to values of k_∞, skin-friction, and $w = sf^2$,

(iii) evaluation of η along the path by quadrature of (1.18-58).

Problem 1.18-1. Show that the nonlinear diffusion equation

$$\frac{\partial c}{\partial t} = D \frac{\partial}{\partial x}\left(\frac{c}{c_0}\frac{\partial c}{\partial x}\right)$$

with the boundary conditions

$$c(x,0) = 0 \quad x > 0$$

$$c(0,t) = c_0 \quad t > 0$$

has a similarity solution of the form

$$\frac{c}{c_0} = g(\xi), \quad \xi = \frac{x}{\sqrt{2Dt}}$$

and derive the ordinary differential equation and boundary conditions for g. Show that this second-order equation admits a group and thus reduce the system to the study of a first-order equation plus a quadrature. Show that this first-order equation is essentially the

Blasius equation (1.18-64) (why?) and discuss which path gives the solution. Describe the concentration profile.

Example 1.18-4. Shallow Membrane Equations

In a geometrically nonlinear theory of axi-symmetric deformation of a membrane under pressure loading the following equation is obtained:[1]

$$\frac{d}{dx}\left(x^3 \frac{dy}{dx}\right) = x^3\nu(x) - \frac{x^3}{y^2} q(x) \qquad 0 < x < 1. \tag{1.18-72}$$

Here y = deflection from original shape

x = radial coordinate

$\nu(x)$ = shape function, ν = const. for sphere

$q(x)$ = load function, $q \sim x^4$ for uniform pressure; all of these quantities have been made dimensionless. The boundary conditions are

$$\begin{aligned} y(0) &= \text{finite} \quad &&\text{(regularity at axis)} \\ y(1) &= 0 &&\text{(fixed at edge)} \end{aligned} \tag{1.18-73}$$

It is interesting to search for those shapes $\nu(x)$ and loadings $q(x)$ such that the basic equations can be reduced in order and a phase plane studied. In this way the existence of the solution can be shown and its qualitative features can be elucidated.

In this section we search for those groups and functions $\nu(x)$, $q(x)$ which leave the equation invariant. The basic equation (1.18-72) can be written in standard form

$$\Omega(x,y,y',y'') = y'' - \omega(x,y,y') = 0 \tag{1.18-74}$$

where

---

[1] Bauer, L., Reiss, E. and Callegari, A., On the Collapse of Shallow Elastic Membranes, Proc. Symp. on Nonlinear Elasticity, University of Wisconsin, 1973, to appear.

$$\omega(x,y,y') = \nu(x) - \frac{q(x)}{y^2} - \frac{3}{x}y' .$$

Note:

$$\omega_x = \nu' - \frac{q'}{y^2} + \frac{3}{x^2}y', \quad \omega_y = 2\frac{q}{y^3}, \quad \omega_{y'} = -\frac{3}{x} .$$

The general condition for group invariance (1.16-19) is

$$\begin{aligned}\xi\omega_x + \eta\omega_y + \{\eta_x + (\eta_y - \xi_x)y' - \xi_y y'^2\}\omega_{y'} - \eta_{xx} \\ -(2\eta_{xy}-\xi_{xx})y' - (\eta_{yy}-2\xi_{xy})y'^2 + \xi_{yy}y'^3 \\ - (\eta_y - 2\xi_x - 3\xi_y y')\omega = 0 \quad \text{for all} \quad x,y,y' .\end{aligned} \tag{1.18-75}$$

In order to derive the conditions on (ξ,η), the coefficients of different powers of y' are set to zero; further in the resulting expressions the functions of x which are coefficients of certain powers of y are also set to zero. This treatment is necessary since the original conditions (1.18-75) must be satisfied identically for all (x,y,y'). For example:

coeff. of y'^3: $\xi_{yy} = 0$

$$\xi(x,y) = A(x) + yB(x). \tag{1.18-76}$$

coeff. of y'^2: $\eta_{yy} = 2B' - \frac{6}{x}B$

$$\eta(x,y) = F(x) + yG(x) + y^2\left[B' - \frac{3}{x}B\right]. \tag{1.18-77}$$

coeff. of y' : coeff. of $\frac{1}{y^2}$ ==> $B = 0$ (1.18-78)

also

coeff. of $\frac{1}{y^3}$ in (1.18-75): ==> $F = 0$ (1.18-79) coeff. of y' now reads:

$$2G' = A'' - 3\left(\frac{A}{x}\right)' \tag{1.18-80}$$

so that

$$2G(x) = A' - \frac{3}{x}A + C_1. \tag{1.18-81}$$

For the remaining terms (a polynomial in y with coefficients depending on x), we get:

coeff. of y^0: $-A\nu' + (G-2A')\nu = 0$

or

$$\frac{\nu'}{\nu} = \frac{G - 2A'}{A}. \tag{1.18-82}$$

coeff. of y: $\frac{3}{x}G' + G'' = 0$

or

$$G(x) = C_3 - \frac{1}{2}\frac{C_2}{x^2}. \tag{1.18-83}$$

coeff. of y^{-2}: $Aq' - 2Gq - (G-2A')q = 0$

or

$$\frac{q'}{q} = \frac{3G - 2A'}{A}. \tag{1.18-84}$$

Now integration of (1.18-81) using (1.18-83) for G gives

$$A(x) = \frac{1}{4}\frac{C_2}{x} + C_4x + C_5x^3.$$

Then, (1.18-82, 84) become

$$\begin{aligned} \frac{\nu'}{\nu} &= \frac{(C_3-2C_4) - 6C_5x^2}{C_4x + C_5x^3 + \frac{1}{4}\frac{C_2}{x}} \\ \frac{q'}{q} &= \frac{(3C_3-2C_4) - \frac{C_2}{x^2} - 6C_5x^2}{C_4x + C_5x^3 + \frac{1}{4}\frac{C_2}{x}} \end{aligned} \tag{1.18-85}$$

$C_{2,3,4,5}$ are arbitrary constants of the group. For regularity as $x \to 0$ it is necessary that $C_2 = 0$ (if not, $q \to \infty$). We can choose $C_4 = 1$ and let $C_5 = \alpha$, $C_3 = \beta$. Then we obtain

$$\begin{aligned} A(x) &= x + \alpha x^3 \\ G(x) &= \beta \\ \frac{\nu}{\nu_o} &= \frac{x^{\beta-2}}{(1+\alpha x^2)^{\frac{1}{2}\beta+2}} \\ \frac{q}{q_o} &= \frac{x^{3\beta-2}}{(1+\alpha x^2)^{\frac{3}{2}\beta+2}} . \end{aligned} \tag{1.18-86}$$

We consider further only the case $\beta = 2$ where the membrane is spherical and loaded with constant pressure near $x = 0$.

Summarizing, the differential equation (1.18-72) is

$$\frac{d}{dx}\left(x^3 \frac{dy}{dx}\right) = \nu_o \frac{x^3}{(1+\alpha x^2)^3} - q_o \frac{x^7}{(1+\alpha x^2)^5 y^2} \tag{1.18-87}$$

and this equation is invariant under the group given by

$$\begin{aligned} \xi(x,y) &= x + \alpha x^3 \\ \eta(x,y) &= 2y \end{aligned} . \tag{1.18-88}$$

This equation (1.18-87) can be reduced to a first-order equation by finding two invariants of the group (u,v).

These are found from solving the characteristic differential equations:

$$\frac{dx}{x + \alpha x^3} = \frac{dy}{2y} = \frac{d(y')}{[2 - (1+3\alpha x^2)](y')} . \tag{1.18-89}$$

Integration of the first two of (1.18-89) gives as invariant $u = \text{const.}$, namely,

$$u(x,y) = \frac{y(1+\alpha x^2)}{x^2} \tag{1.18-90}$$

while the integration of the first and third gives

$$v(x,y,y') = \frac{y'}{x}(1+\alpha x^2)^2 . \tag{1.18-91}$$

A first-order differential equation for $v(u)$ can be found directly as follows:

$$\frac{du}{dx} = \frac{y'}{x^2} - 2\frac{y}{x^3} + \alpha y' = \frac{1}{x(1+\alpha x^2)}(v-2u) \tag{1.18-92}$$

$$\frac{dv}{dx} = \frac{y''(1+\alpha x^2)^2}{x} - \frac{y'(1+\alpha x^2)^2}{x^2} + 4\alpha(1+\alpha x^2)y' . \tag{1.18-93}$$

(1.18-92) provides a mapping back to the x-coordinate along a trajectory $v(u)$

$$\frac{dx}{x(1+\alpha x^2)} = \frac{du}{v-2u} . \tag{1.18-94}$$

The expression for y'' from (1.18-93) can be substituted in the original equation (1.18-87) written as

$$y'' + \frac{3}{x}y' = \frac{1}{(1+\alpha x^2)^3}\left\{\nu_o - \frac{q_o}{u^2}\right\} \tag{1.18-95}$$

to yield an equation for $\frac{dv}{du}$. Note

$$y'' + \frac{3}{x}y' = \frac{x}{(1+\alpha x^2)^2}\frac{dv}{dx} + \frac{y'}{x} - \frac{4\alpha x y'}{1+\alpha x^2} + \frac{3}{x}y'$$

$$= \frac{1}{(1+\alpha x^2)^3}\left\{(v-2u)\frac{dv}{du} + 4v\right\}$$

on using (1.18-91, 94). Thus, (1.18-95) becomes

$$(v-2u)\frac{dv}{du} + 4v = \nu_o - \frac{q_o}{u^2} \tag{1.18-96}$$

or

$$\frac{dv}{du} = \frac{\nu_o u^2 - q_o - 4u^2 v}{u^2(v-2u)} . \tag{1.18-97}$$

The paths of (1.18-97) in the v,u-plane can be studied and the particular path representing the solution satisfying the boundary conditions can be isolated.

Exercise: Show that for β arbitrary

$$\frac{dv}{du} = \frac{\nu_o u^2 - q_o - (\beta+2)u^2 v}{u^2(v-\beta u)}$$

$$\frac{du}{v - \beta u} = \frac{dx}{x(1+\alpha x^2)}$$

$$u = \frac{y}{x^\beta}(1+\alpha x^2)^{\beta/2}$$

$$v = \frac{y'}{x^{\beta-1}}(1+\alpha x^2)^{\frac{\beta}{2}+1} .$$

2. PARTIAL DIFFERENTIAL EQUATIONS

2.0. Partial Differential Equations.

The role of Lie theory in constructing solutions to partial differential equations differs essentially from its role for solving ordinary differential equations.[1] Invariance under a one-parameter Lie group of transformations reduces by one the number of variables appearing in a partial differential equation rather than the order as is the case for an ordinary differential equation. The method is not an encompassing as in Part 1 since it leads to particular (similarity) solutions and not to the "general" solution of a given partial differential equation. Thus boundary conditions play an important role in the applications of Lie theory to partial differential equations.

As in Part 1 the equivalence of a Lie group of transformations to a translation group in terms of canonical coordinates is most important to a basic understanding. For a second order partial differential equation of first degree in one of the second derivatives note the difference between the general case and the case when one of the independent variables is missing, i.e.,

$$u_{xx} = H(u_{xt}, u_{tt}, u_x, u_t, u, x, t) \text{ , general} \qquad (2.0\text{-}1)$$

and

$$u_{xx} = H(u_{xt}, u_{tt}, u_x, u_t, u, x) \text{ , } t \text{ missing} \qquad (2.0\text{-}2)$$

In the latter case (provided (2.0- 2) is well-behaved) there will exist particular solutions $u = F(x)$ where $F(x)$ satisfies the second order ordinary differential equation

$$F''(x) = H(0, 0, F'(x), 0, F(x) \qquad (2.0\text{-}3)$$

---

(1) We assume that the given partial differential equation is not of first order since such an equation is equivalent to a characteristic system of ordinary differential equations to which the methods of Part 1 apply.

As another example consider the wave equation

$$u_{xx} = u_{tt} \quad . \tag{2.0-4}$$

Clearly the one-parameter (α) group of transformations

$$\begin{aligned} x^* &= \alpha x \\ t^* &= \alpha t \end{aligned} \tag{2.0-5}$$

leaves invariant (2.0-4). Choosing canonical coordinates $r = \frac{x}{t}$, $s = \log t$, (2.0-4) becomes

$$(1 - r^2)\, u_{rr} - u_{ss} + 2r\, u_{rs} + u_s - 2r\, u_r = 0 \tag{2.0-6}$$

so that $u = F(r)$ is a particular solution of (2.0-4) provided

$$(1 - r^2)\, F'' - 2rF' = 0 \quad . \tag{2.0-7}$$

We immediately see that it is unnecessary to derive (2.0-6). In order to generate the particular solution form $u = F(r)$ the important point is the invariance of (2.0-4) under the one-parameter (ε) Lie group of transformations

$$\begin{aligned} r^* &= r \\ s^* &= s + \varepsilon \quad . \end{aligned}$$

In general, when reducing the number of variables of a partial differential equation from symmetry considerations, the invariants of the group become the new variables. The generated similarity solution satisfies an auxiliary first order partial differential equation (called the <u>invariant surface condition</u>) whose coefficients depend on the infinitesimals of the group. Solving the corresponding characteristic equations of this first order p.d.e. we find the functional form of the similarity solution. For a partial differential equation with independent variables (x, t) and dependent variable u typically one of the invariants will be of the form $\zeta(x, t)$ and the other invariant can be expressed as an arbitrary function of ζ, $F(\zeta)$. The functional form for the similarity solution will be

$$u = \mathscr{F}(x, t, \zeta, F(\zeta)) \qquad (2.0\text{-}8)$$

ζ is called the similarity variable and $F(\zeta)$ becomes the new dependent variable. The dependence of $\mathscr{F}$ on $\{x, t, \zeta, F(\zeta)\}$ is known explicitly and by substituting (2.0-8) into the given partial differential equation we obtain an ordinary differential equation for $F(\zeta)$.

Thus for a given problem we first seek the largest set of infinitesimals leaving invariant the governing partial differential equation (s) . The infinitesimals satisfy a set of "determining equations" which are of such a number that they always seem to be solvable in closed form.[2] Next we analyse the symmetries of the boundary conditions and seek the subgroup (actually the subalgebra of the infinitesimals) leaving invariant the boundary curves and the boundary conditions prescribed on them. For linear partial differential equations it is unnecessary to leave invariant all of the boundary conditions.

The use of group invariance to reduce the number of variables appearing in a partial differential equation does not depend on the governing p.d.e. being linear. The transformations considered are functions of both independent and dependent variables. It will be seen that without consideration of infinitesimals it is often impossible to recognize the complete symmetry of a given partial differential equation.

---

(2) The solvability in closed form of the determining equations comes from the fact that they form an overdetermined system of equations. The determining equations are always linear.

In §2.5 it is shown that the use of dimensional analysis to reduce the number of variables appearing in a partial differential equation is a special case of invariance under a group of stretchings. §2.6 discusses how group invariance leads to new solutions from known solutions of a given partial differential equation. Here the distinction between linear and nonlinear equations is significant. In §2.7 the symmetry properties of the one-dimensional heat equation are discussed in detail. This partial differential equation is shown to be invariant under a six-parameter Lie group. Two of these parameters cannot be spotted by inspection, i.e., without using general procedures to find the infinitesimals.

In §2.8 the group properties of the heat equation are exploited in considering various boundary value problems. In §2.8 the use of complex parameters is exemplified and in this section as well as in §2.9 and in §2.10 invariance under a multi-parameter group is discussed and applied to a number of examples.

The use of an infinite-parameter group to derive the Poisson kernel is given in §2.12. Nonlinear examples are treated in §2.13 and §2.14. An example in §2.14 illustrates the use of group invariance to convert a boundary value problem into an initial value problem.

The works of Lie(3), Ovsjannikov(4) and Matschat and Müller(5)

---

(3) S. Lie, Über die Integration durch bestimmte Integrale von einer Klasse linearer partieller Differentialgleichungen, Arch. for Math., Vol. VI, No. 3, Kristiana, 1881, p. 328.

(4) L. V. Ovsjannikov, Gruppovye svoystva differentsialny uravneni, Novosibirsk, 1962. (Group properties of differential equations - translated by G. Bluman, 1967).

(5) E. A. Müller and K. Matschat, Über das Auffinden von Ähnlichkeitslösungen partieller Differentialgleichungssysteme unter Benutzung von Transformationsgruppen, mit Anwendungen auf Probleme der Stromungsphysik, Miszellaneen der Angewandten Mechanik, Berlin, 1962, p. 190

are highly recommended. Lie essentially found the group of the heat equation but he did not seem to realise that group invariance could be used to reduce the number of variables appearing in a partial differential equation. Ovsjannikov gives a systematic study of the use and derivation of Lie groups and infinitesimal transformations to reduce the number of variables of a partial differential equation. Among his accomplishments Ovsjannikov derives the group properties of the equation of one-dimensional adiabatic gas dynamics and nonlinear diffusion. However he does not consider applications to boundary value problems. Matschat and Müller consider the symmetry properties of systems of quasilinear partial differential equations of first order under contact transformations.

2.1. Formulation of Invariance for the Special Case of One Dependent and Two Independent Variables.

In this section we restrict ourselves to a system of equations consisting of a single second order partial differential equation and associated boundary conditions. It will be shown that if such a system is invariant under a one-parameter Lie group of transformations then the partial differential equation is reduced to an ordinary differential equation. The following points should come to light:

(i) As in reducing the order of an ordinary differential equation in Part 1 only the infinitesimal of a group appears to be needed to reduce the number of variables of a partial differential equation. This procedure generates a similarity form for the solution.

(ii) The order of the partial differential equation is irrelevant.

(iii) The given partial differential equation can be nonlinear.

(iv) The transformations considered are functions of both dependent and independent variables.

(v) Boundary conditions are unnecessary to apply group methods to reduce the number of variables of a partial differential equation where the aim is to find a class of solutions rather than to solve a particular problem.

(vi) The infinitesimal of the group transforming the variables of a partial differential equation can be naturally extended to the infinitesimals of the derivatives of the dependent variables.

(vii) By "thinking infinitesimally" we can find the group of transformations leaving invariant a given partial differential equation. The determining equations for the infinitesimals are linear. As shown before in Part 1 the infinitesimals generate the group.

In what follows u_x denotes $\frac{\partial u}{\partial x}$, u_t denotes $\frac{\partial u}{\partial t}$, etc.

Consider the partial differential equation (dependent variable u, independent variables x and t)

$$H(u_{xx}, u_{xt}, u_{tt}, u_x, u_t, u, x, t) = 0 \tag{2.1-1}$$

with m associated boundary conditions

$$B_\beta(u_x, u_t, u, x, t) = 0 \tag{2.1-2}$$

on prescribed curves

$$w_\beta(x, t) = 0\,, \qquad \beta = 1, 2, \ldots, m\,. \tag{2.1-3}$$

(There may be no associated boundary conditions) (2.1-1, 2, 3) will be denoted by system S.

Consider a one-parameter (ε) Lie group of transformations:

$$u^* = U^*(x, t, u; \varepsilon)$$

$$x^* = X^*(x, t, u; \varepsilon) \tag{2.1-4}$$

$$t^* = T^*(x, t, u; \varepsilon) \quad .$$

Say $u = \Theta(x, t)$ is a solution[(1)] to system S .

Consider the system S^* obtained from S by having u replaced by v ,

x by $x^* = X^*(x, t, \Theta(x, t); \varepsilon)$,

t by $t^* = T^*(x, t, \Theta(x, t); \varepsilon)$.

A solution of system S^* is $v = \Theta(x^*, t^*)$.

We are now ready to define <u>invariance</u>.

<u>Definition 2.1-1.</u>

(2.1-4) leaves S <u>invariant</u> iff $v = U^*(x, t, \Theta(x, t); \varepsilon)$ satisfies S^* whenever $u = \Theta(x, t)$ is a solution to S .

From invariance there are two candidates for a solution to S^* , namely

$$v = v_1 = \Theta(x^*, t^*)$$

and

$$v = v_2 = U^*(x, t, \Theta(x, t); \varepsilon) \quad .$$

If we assume that S (hence S^*) has a unique solution then

$$v_1 \equiv v_2 \quad .$$

Hence $\Theta(x, t)$ must satisfy the one-parameter (ε) functional equation

---

(1) By solution here we mean that substitution of $u = \Theta(x, t)$ in (2.1-1,2,3) satisfies these relations identically in (x, t) in some suitable domain.

$$\Theta(X^*(x, t, \Theta(x, t); \varepsilon), T^*(x, t, \Theta(x, t); \varepsilon)) = U^*(x, t, \Theta(x, t); \varepsilon) \quad . \qquad (2.1\text{-}5)^{(2)}$$

Our objective now is to find $\Theta(x, t)$ given that (2.1-4) leaves S invariant. For the time being we put aside the question of how to find (2.1-4).

Expanding (2.1-4) about the identity $\varepsilon = 0$ we generate the following infinitesimal $(\eta, \xi, \tau)(0(\varepsilon))$ terms:

$$\begin{aligned} u^* &= u + \varepsilon\, \eta(x, t, u) + 0(\varepsilon^2) \\ x^* &= x + \varepsilon\, \xi(x, t, u) + 0(\varepsilon^2) \\ t^* &= t + \varepsilon\, \tau(x, t, u) + 0(\varepsilon^2) \quad . \end{aligned} \qquad (2.1\text{-}6)$$

We now use (2.1-6) to expand the functional equation (2.1-5) about $\varepsilon = 0$:

$$\begin{aligned} &\Theta(x + \varepsilon\, \xi(x, t, \Theta) + 0(\varepsilon^2), t + \varepsilon\, \tau(x, t, \Theta) + 0(\varepsilon^2)) \\ &\qquad = \Theta(x, t) + \varepsilon\, \eta(x, t, \Theta) + 0(\varepsilon^2) \quad . \end{aligned} \qquad (2.1\text{-}7)$$

The $0(\varepsilon)$ term of the functional equation (2.1-7) leads to the following first order partial differential equation satisfied by $\Theta(x, t)$, given (η, ξ, τ) :

$$\xi(x, t, \Theta)\Theta_x + \tau(x, t, \Theta)\Theta_t = \eta(x, t, \Theta) \quad . \qquad (2.1\text{-}8)$$

(2.1-8) is called the <u>invariant surface condition</u>. The method of solving equations of the form (2.1-8) is discussed in the Appendix. In particular the general solution of (2.1-8) is a family of surfaces.

---

(2) It could happen that S does not have a unique solution due to the lack of a sufficient number of boundary conditions. In this case we could augment S with further boundary conditions in such a way that the augmented system is still invariant under (2.1-4) and also has a unique solution. If S does not have any associated boundary conditions then (2.1-4) is a group of transformations leaving invariant (2.1-1) only. The functional equation (2.1-5), in these cases, leads us to a class of solutions of (2.1-1).

If $F(x, t, \Theta) = 0$ defines a surface satisfying (2.1-8) then this surface is an invariant of the group (2.1-4) since

$$\xi(x, t, \Theta)F_x + \tau(x, t, \Theta)F_t + \eta(x, t, \Theta)F_\Theta = 0$$

(cf. discussion of invariant surfaces in §1.12).

The general solution of (2.1-8) is obtained by solving the characteristic equations

$$\frac{dx}{\xi(x, t, \Theta)} = \frac{dt}{\tau(x, t, \Theta)} = \frac{d\Theta}{\eta(x, t, \Theta)} \tag{2.1-9}$$

In principle, the general solution of (2.1-9) can be found. It involves two constants, one of which becomes the independent variable $\zeta(x, t, \Theta)$, called the <u>similarity variable</u>, and the other plays the role of a dependent variable, $f(\zeta)$. We obtain the similarity form

$$\Theta = \mathscr{F}(x, t, f(\zeta)) \tag{2.1-10}$$

with the dependence of $\mathscr{F}$ on x, t and the arbitrary function $f(\zeta)$ known explicitly. Substitution of (2.1-10) into (2.1-1) results in an ordinary differential equation for $f(\zeta)$. $\frac{\xi}{\tau} = \text{fn}(x, t)$ is a commonly occurring case. Here

$$\zeta(x, t) = \text{constant } (\Rightarrow x = g(\zeta, t)) \tag{2.1-11}$$

would be the integral of the first equality in (2.1-9) and defines path curves (<u>similarity curves</u>) in (x, t)-space. Then the second equation in (2.1-9) becomes

$$\frac{dt}{\tau(g(\zeta, t), t, \Theta)} = \frac{d\Theta}{\eta(g(\zeta, t), t, \Theta)} \tag{2.1-12}$$

with solution

$$G(t, \zeta(x, t), \Theta) = \text{const.} = f(\zeta) \quad . \tag{2.1-13}$$

Hence

$$\Theta = \mathscr{F}(x, t, f(\zeta(x, t))) \quad . \tag{2.1-14}$$

Problem 2.1-1

Show that if (2.1-4) is such that X^* and T^* are independent of u and U^* depends linearly on u then the general solution of (2.1-8) is of the form

$$\Theta = f(\zeta)\mathcal{F}(x, t) \tag{2.1-15}$$

where f is an arbitrary function of ζ, ζ is a known function of x and t, and $\mathcal{F}$ is a known function of x and t.

So far we have shown that given a group of transformations (2.1-4) leaving invariant (2.1-1) we use the infinitesimal (2.1-6) of the group to reduce (2.1-1) to an ordinary differential equation. We now turn our attention to the problem of finding the group of transformations leaving invariant a partial differential equation. Since it is the infinitesimal of a group which is used to reduce the number of variables appearing in a partial differential equation we seek the infinitesimal (2.1-6) leaving invariant (2.1-1)[3]. In order to find the infinitesimal we need to extend the group to calculate how derivative terms transform. The method of extension is the same as in Part 1 for ordinary derivatives. In what follows $u = \Theta(x, t)$, $u^* = U^*(x, t, u; \varepsilon)$, i.e., $u = \Theta(x, t)$ will be some solution to S with $v = U^*(x, t, \Theta(x, t); \varepsilon)$ the corresponding solution to S^*. First extensions refer to how the first partial derivatives transform; second extensions refer to second partial derivatives. Before calculating these extensions we need to calculate the auxiliary functions

(3) If boundary conditions are present we first seek a group leaving invariant the governing partial differential equation and then find a subgroup (hopefully not trivial) leaving invariant the boundary conditions and the curves on which they are prescribed. Examples in §2.8 will illustrate the use of boundary conditions.

$\frac{\partial x}{\partial x^*}$, $\frac{\partial x}{\partial t^*}$, $\frac{\partial t}{\partial t^*}$ and $\frac{\partial t}{\partial x^*}$. By $\frac{\partial x}{\partial x^*}$ we understand that $u = \Theta(x,t)$ and that only t^* is held fixed. Hence

$$\frac{\partial x}{\partial x^*} = \frac{\partial}{\partial x^*}[x^* - \varepsilon\,\xi(x,t,\Theta) + 0(\varepsilon^2)]$$

$$= 1 - \varepsilon\left[\frac{\partial \xi}{\partial x}\frac{\partial x}{\partial x^*} + \frac{\partial \xi}{\partial u}\frac{\partial \Theta}{\partial x}\frac{\partial x}{\partial x^*}\right] + 0(\varepsilon^2)$$

$$= 1 - \varepsilon\left[\frac{\partial \xi}{\partial x} + \frac{\partial \xi}{\partial u}\Theta_x\right] + 0(\varepsilon^2) \quad .$$

Similarly

$$\frac{\partial x}{\partial t^*} = -\varepsilon\left[\frac{\partial \xi}{\partial t} + \frac{\partial \xi}{\partial u}\Theta_t\right] + 0(\varepsilon^2)$$

$$\frac{\partial t}{\partial t^*} = 1 - \varepsilon\left[\frac{\partial \tau}{\partial t} + \frac{\partial \tau}{\partial u}\Theta_t\right] + 0(\varepsilon^2) \tag{2.1-16}$$

$$\frac{\partial t}{\partial x^*} = -\varepsilon\left[\frac{\partial \tau}{\partial x} + \frac{\partial \tau}{\partial u}\Theta_x\right] + 0(\varepsilon^2) \quad .$$

Recall that

$$u^* = U^*(x,t,\Theta(x,t);\varepsilon) = \Theta(x,t) + \varepsilon\eta(x,t,\Theta) + 0(\varepsilon^2) \quad .$$

Hence we find that the extensions are:

<u>First Extensions</u> :

$$\begin{aligned}\frac{\partial u^*}{\partial x^*} &= \frac{\partial[\Theta(x,t) + \varepsilon\eta(x,t,\Theta)]}{\partial x^*} + 0(\varepsilon^2) \\ &= \frac{\partial[\Theta(x,t) + \varepsilon\eta(x,t,\Theta)]}{\partial x}\frac{\partial x}{\partial x^*} + \frac{\partial t}{\partial x^*}\Theta_t + 0(\varepsilon^2) \quad .\end{aligned} \tag{2.1-17}$$

Substituting (2.1-16) into (2.1-17) we are led to

$$\frac{\partial u^*}{\partial x^*} = \Theta_x + \varepsilon\left[\frac{\partial \eta}{\partial x} + \left(\frac{\partial \eta}{\partial u} - \frac{\partial \xi}{\partial x}\right)\Theta_x - \frac{\partial \tau}{\partial x}\Theta_t - \frac{\partial \xi}{\partial u}\Theta_x^2 - \frac{\partial \tau}{\partial u}\Theta_x\Theta_t\right] + 0(\varepsilon^2) \quad . \tag{2.1-18}$$

Let η_x, η_t denote the infinitesimals of $\frac{\partial u^*}{\partial x^*}$ and $\frac{\partial u^*}{\partial t^*}$ respectively.[4] Then

$$\eta_x = \frac{\partial \eta}{\partial x} + \left[\frac{\partial \eta}{\partial u} - \frac{\partial \xi}{\partial x}\right]\Theta_x - \frac{\partial \tau}{\partial x}\Theta_t - \frac{\partial \xi}{\partial u}\Theta_x^2 - \frac{\partial \tau}{\partial u}\Theta_x\Theta_t \tag{2.1-19}$$

and similarly

$$\eta_t = \frac{\partial \eta}{\partial t} + \left[\frac{\partial \eta}{\partial u} - \frac{\partial \tau}{\partial t}\right]\Theta_t - \frac{\partial \xi}{\partial t}\Theta_x - \frac{\partial \tau}{\partial u}\Theta_t^2 - \frac{\partial \xi}{\partial u}\Theta_x\Theta_t \; . \tag{2.1-20}$$

Second Extensions:

The details of calculations for the second extensions, η_{xx}, η_{tt} and η_{xt} are left to the reader. A general discussion for calculating the infinitesimals of extended transformations will be given in §2.2. The results are:

$$\eta_{xx} = \frac{\partial^2 \eta}{\partial x^2} + \left[2\frac{\partial^2 \eta}{\partial x \partial u} - \frac{\partial^2 \xi}{\partial x^2}\right]\Theta_x - \frac{\partial^2 \tau}{\partial x^2}\Theta_t + \left[\frac{\partial^2 \eta}{\partial u^2} - 2\frac{\partial^2 \xi}{\partial x \partial u}\right]\Theta_x^2$$

$$- 2\frac{\partial^2 \tau}{\partial x \partial u}\Theta_x\Theta_t - \frac{\partial^2 \xi}{\partial u^2}\Theta_x^3 - \frac{\partial^2 \tau}{\partial u^2}\Theta_x^2\Theta_t + \left[\frac{\partial \eta}{\partial u} - 2\frac{\partial \xi}{\partial x}\right]\Theta_{xx} - 2\frac{\partial \tau}{\partial x}\Theta_{xt}$$

$$- 3\frac{\partial \xi}{\partial u}\Theta_{xx}\Theta_x - \frac{\partial \tau}{\partial u}\Theta_{xx}\Theta_t - 2\frac{\partial \tau}{\partial u}\Theta_{xt}\Theta_x \tag{2.1-21}$$

$$\eta_{tt} = \frac{\partial^2 \eta}{\partial t^2} + \left[2\frac{\partial^2 \eta}{\partial t \partial u} - \frac{\partial^2 \tau}{\partial t^2}\right]\Theta_t - \frac{\partial^2 \xi}{\partial t^2}\Theta_x + \left[\frac{\partial^2 \eta}{\partial u^2} - 2\frac{\partial^2 \tau}{\partial t \partial u}\right]\Theta_t^2 - 2\frac{\partial^2 \xi}{\partial t \partial u}\Theta_x\Theta_t$$

$$- \frac{\partial^2 \tau}{\partial u^2}\Theta_t^3 - \frac{\partial^2 \xi}{\partial u^2}\Theta_t^2\Theta_x + \left[\frac{\partial \eta}{\partial u} - 2\frac{\partial \tau}{\partial t}\right]\Theta_{tt} - 2\frac{\partial \xi}{\partial t}\Theta_{xt} - 3\frac{\partial \tau}{\partial u}\Theta_{tt}\Theta_t$$

$$- \frac{\partial \xi}{\partial u}\Theta_{tt}\Theta_x - 2\frac{\partial \xi}{\partial u}\Theta_{xt}\Theta_t \tag{2.1-22}$$

---

(4) It is easy to show that in this manner the once extended group is a group of transformations of the variables $\{x,t,u,u_x,u_t\}$ provided that the "unextended" group is a group of transformations of the variables $\{x,t,u\}$. u_x refers to Θ_x; u_t to Θ_t.

$$\eta_{xt} = \frac{\partial^2\eta}{\partial x\partial t} + \left[\frac{\partial^2\eta}{\partial x\partial u} - \frac{\partial^2\tau}{\partial t\partial x}\right]\Theta_t + \left[\frac{\partial^2\eta}{\partial t\partial u} - \frac{\partial^2\xi}{\partial t\partial x}\right]\Theta_x - \frac{\partial^2\tau}{\partial x\partial u}\Theta_t^2$$

$$+ \left[\frac{\partial^2\eta}{\partial u^2} - \frac{\partial^2\xi}{\partial x\partial u} - \frac{\partial^2\tau}{\partial u\partial t}\right]\Theta_x\Theta_t - \frac{\partial^2\xi}{\partial t\partial u}\Theta_x^2 - \frac{\partial^2\tau}{\partial u^2}\Theta_x\Theta_t^2 - \frac{\partial^2\xi}{\partial u^2}\Theta_t\Theta_x^2$$

$$- \frac{\partial\tau}{\partial x}\Theta_{tt} + \left[\frac{\partial\eta}{\partial u} - \frac{\partial\xi}{\partial x} - \frac{\partial\tau}{\partial t}\right]\Theta_{xt} - \frac{\partial\xi}{\partial t}\Theta_{xx} - 2\frac{\partial\tau}{\partial u}\Theta_t\Theta_{xt}$$

$$- 2\frac{\partial\xi}{\partial u}\Theta_x\Theta_{xt} - \frac{\partial\tau}{\partial u}\Theta_x\Theta_{tt} - \frac{\partial\xi}{\partial u}\Theta_t\Theta_{xx} \quad . \qquad (2.1\text{-}23)$$

For invariance of (2.1-1) under the group of transformations (2.1-4) it is necessary that for any solution $u = \Theta(x, t)$ of (2.1-1),

$$H(u^*_{x^*x^*}, u^*_{x^*t^*}, u^*_{t^*t^*}, u^*_{x^*}, u^*_{t^*}, u^*, x^*, t^*) = 0 \qquad (2.1\text{-}24)$$

i.e. the partial differential equation (2.1-1) is an invariant of the twice extended group (2.1-4). A necessary and sufficient condition for (2.1-1) to be an invariant of (2.1-4) is that the infinitesimal $(O(\varepsilon))$ term of (2.1-24) is identically zero. We are now thinking of (2.1-1) as an equation in 8 variables each of which has its own infinitesimal element for any group of transformations (2.1-4). Hence (2.1-4) <u>leaves invariant</u> (2.1-1) iff given any $u = \Theta(x, t)$ satisfying (2.1-1), $\Theta(x, t)$ also satisfies the partial differential equation (now containing the to be determined infinitesimal elements η, ξ, and τ):

$$\eta_{xx}\frac{\partial H}{\partial u_{xx}} + \eta_{xt}\frac{\partial H}{\partial u_{xt}} + \eta_{tt}\frac{\partial H}{\partial u_{tt}} + \eta_x\frac{\partial H}{\partial u_x} + \eta_t\frac{\partial H}{\partial u_t}$$

$$+ \eta\frac{\partial H}{\partial u} + \xi\frac{\partial H}{\partial x} + \tau\frac{\partial H}{\partial t} = 0 \quad . \qquad (2.1\text{-}25)$$

In order to find the infinitesimal elements $(\eta, \varepsilon, \tau)$ leaving invariant (2.1-1) we first substitute the partial differential equation (2.1-1) into (2.1-25)[5]. The resulting equation is treated as a form in the derivatives of Θ whose coefficients depend on (Θ, x, t) and the unknowns (η, ξ, τ). After the substitution we collect together the coefficients of like derivative terms in Θ and set all of them equal to zero. The resulting equations are called the determining equations of the group. In practice these equations are solvable and thus (η, ξ, τ) are determined.

Problem 2.1-2

Show that the determining equations for (η, ξ, τ) are linear and homogeneous in these unknowns for any polynomial partial differential equation. How many determining equations are there for an arbitrary linear homogeneous second order partial differential equation?

Without loss of continuity the reader could now skip to §2.7 where the results of this section are applied to finding the general similarity solution of the heat equation.

2.2. Formulation of Invariance in General

In this section we will generalize the results of §2.1 to systems of partial differential equations containing an arbitrary number of dependent and independent variables. It will be shown how to calculate in general the infinitesimal of the extended transformation and under what conditions a one-parameter group of transformations leaving invariant a system of partial differential equations reduces

---

[5] We assume that in (2.1-1) derivatives of u appear in polynomial form. Such a partial differential equation will be called a polynomial partial differential equation.

the number of independent variables by one. In what follows a repeated index denotes summation over the index, superscripts and subscripts refer to dependent and independent variables respectively. A subscript preceded by a comma will refer to a partial derivative with respect to the variable represented by the subscript. We will consider systems of partial differential equations containing m dependent and n independent variables. Let

$$\begin{aligned} &u^i, \quad i = 1, 2, \ldots, m \\ &x_j, \quad j = 1, 2, \ldots, n \end{aligned} \tag{2.2-1}$$

represent dependent and independent variables respectively.

Consider a one-parameter (ε) group of transformations

$$\begin{aligned} u^{*i} &= U^{*i}(x_1, x_2, \ldots, x_n, u^1, u^2, \ldots, u^m; \varepsilon) \\ x_j^* &= X_j^*(x_1, x_2, \ldots, x_n, u^1, u^2, \ldots, u^m; \varepsilon) \end{aligned} \tag{2.2-2}$$

with corresponding infinitesimal elements $\eta^i(x_1, x_2, \ldots, x_n, u^1, u^2, \ldots, u^m)$ and $\xi_j(x_1, x_2, \ldots, x_n, u^1, u^2, \ldots, u^m)$ obtained by expanding (2.2-2) about the identity $\varepsilon = 0$. We assume that the transformations (2.2-2) have the property that if $u^i = \Theta^i(x_1, x_2, \ldots, x_n)$ then we can invert (2.2-2) so that $u^{*i} = \Theta^{*i}(x_1^*, x_2^*, \ldots, x_n^*)$ for some function Θ^{*i}. $u^i{}_{,j}$ and $u^{*i}{}_{,j}$ denote $\frac{\partial u^i}{\partial x_j}$ and $\frac{\partial u^{*i}}{\partial x^*_j}$, respectively. Throughout this section keep in mind the calculations of §2.1 where $m = 1$ and $n = 2$.

Calculation of the Extended Transformation for Several Variables

We will be computing partial derivatives with respect to x_j of functions depending on x_j and u^i where u^i depends on x_j as in §2.1. To shorten the calculations, we introduce the total derivative operators:

$$\frac{D}{Dx_j} = \frac{\partial}{\partial x_j} + u^\mu{}_{,j} \frac{\partial}{\partial u^\mu}$$
$$\frac{D}{Dx_j^*} = \frac{\partial}{\partial x_j^*} + u^{*\mu}{}_{,j} \frac{\partial}{\partial u^{*\mu}} \quad . \tag{2.2-3}$$

Clearly $\frac{Du^i}{Dx_j} = u^i{}_{,j}$, $\frac{Du^{*i}}{Dx_j^*} = u^{*i}{}_{,j}$ and $\frac{Dx_\alpha^*}{Dx_\beta^*} = \frac{Dx_\alpha}{Dx_\beta} = \delta_{\alpha\beta}$, the Kronecker delta symbol.

Next we derive the chain rule for the total derivative operators: Consider

$$\frac{D}{Dx_j^*} = \frac{\partial x_\nu}{\partial x_j^*}\frac{\partial}{\partial x_\nu} + \frac{\partial u^\mu}{\partial x_j^*}\frac{\partial}{\partial u^\mu} + u^{*\mu}{}_{,j}\frac{\partial x_\nu}{\partial u^{*\mu}}\frac{\partial}{\partial x_\nu} + u^{*\lambda}{}_{,j}\frac{\partial u^\mu}{\partial u^{*\lambda}}\frac{\partial}{\partial u^\mu}$$

$$= \frac{Dx_\nu}{Dx_j^*}\frac{\partial}{\partial x_\nu} + \left[\frac{\partial x_\nu}{\partial x_j^*} + u^{*\lambda}{}_{,j}\frac{\partial x_\nu}{\partial u^{*\lambda}}\right] u^\mu{}_{,\nu}\frac{\partial}{\partial u^\mu} \quad .$$

Hence

$$\frac{D}{Dx_j^*} = \frac{Dx_\nu}{Dx_j^*}\frac{D}{Dx_\nu} \quad . \tag{2.2-4}$$

Moreover the matrix $\left(\frac{Dx_i}{Dx_j^*}\right)$ is the inverse of the matrix $\left(\frac{Dx_i^*}{Dx_j}\right)$ which is obtained directly from (2.2-2). We now have the apparatus necessary for computing the infinitesimals of the extended group. The qth extension refers to how the qth derivative transforms.

First Extensions

$$u^{*i}{}_{,j} = \frac{Du^{*i}}{Dx_j^*}$$
$$= \frac{Du^{*i}}{Dx_\nu}\frac{Dx_\nu}{Dx_j^*}$$
$$= \frac{D[u^i + \varepsilon\eta^i + O(\varepsilon^2)]}{Dx_\nu} \cdot \frac{D[x_\nu^* - \varepsilon\xi_\nu + O(\varepsilon^2)]}{Dx_j^*}$$

$$= u^i{,j} + \varepsilon\left[\frac{D\eta^i}{Dx_j} - \frac{D\xi_\nu}{Dx_j} u^i{,\nu}\right] + 0(\varepsilon^2) \quad . \qquad (2.2\text{-}5)$$

Expanding the right side of (2.2-5), by using (2.2-3), we find that the first extensions, η^i_j , of the continuous transformations (2.2-2) are:

$$\eta^i_j = \frac{\partial\eta^i}{\partial x_j} + \frac{\partial\eta^i}{\partial u^\mu} u^\mu{,j} - \frac{\partial\xi_\nu}{\partial x_j} u^i{,\nu} - \frac{\partial\xi_\nu}{\partial u^\mu} u^\mu{,j}\ u^i{,\nu} \qquad (2.2\text{-}6)$$

where $i = 1,2,\ldots,m$, $j = 1,2,\ldots,n$.

Second Extensions

The second extensions refer to how the second order derivatives, $u^i{}_{,jk}$, transform and can be obtained directly by substituting $u^i{,j}$ and, accordingly, η^i_j for η^i in formula (2.2-5). Thus

$$u^{*i}{,jk} = u^i{,jk} + \varepsilon\left[\frac{D\eta^i_j}{Dx_k} - \frac{D\xi_\nu}{Dx_k} u^i{,j\nu}\right] + 0(\varepsilon^2) \quad . \qquad (2.2\text{-}7)$$

Expanding the right side of (2.2-7) and making the appropriate substitutions, we get the second extensions η^i_{jk} :

$$\begin{aligned}
\eta^i_{jk} = {} & \frac{\partial^2\eta^i}{\partial x_j\partial x_k} + \frac{\partial^2\eta^i}{\partial x_j\partial u^\mu} u^\mu{}_{,k} + \frac{\partial^2\eta^i}{\partial x_k\partial u^\mu} u^\mu{}_{,j} \\
& - \frac{\partial^2\xi_\nu}{\partial x_j\partial x_k} u^i{}_{,\nu} + \frac{\partial\eta^i}{\partial u^\mu} u^\mu{}_{,jk} - \frac{\partial\xi_\nu}{\partial x_j} u^i{}_{,k\nu} \\
& - \frac{\partial\xi_\nu}{\partial x_k} u^i{}_{,j\nu} + \frac{\partial^2\eta^i}{\partial u^\lambda\partial u^\mu} u^\lambda{}_{,j}\ u^\mu{}_{,k} \\
& - \frac{\partial^2\xi_\nu}{\partial x_j\partial u^\mu} u^\mu{}_{,k}\ u^i{}_{,\nu} - \frac{\partial^2\xi_\nu}{\partial x_k\partial u^\mu} u^\mu{}_{,j}\ u^i{}_{,\nu} \\
& - \frac{\partial\xi_\nu}{\partial u^\mu} [u^i{}_{,\nu}\ u^\mu{}_{,jk} + u^\mu{}_{,j}\ u^i{}_{,\nu k} + u^\mu{}_{,k}\ u^i{}_{,j\nu}] \\
& - \frac{\partial^2\xi_\nu}{\partial u^\lambda\partial u^\mu} u^\lambda{}_{,k}\ u^\mu{}_{,j}\ u^i{}_{,\nu}
\end{aligned} \qquad (2.2\text{-}8)$$

where $i = 1,2,\ldots,m$; $j,k = 1,2,\ldots,n$.

Similarly to generate the qth extensions, $\eta^i_{j_1 j_2 \dots j_q}$, one "merely" applies the total differential operator $\frac{D}{Dx_{j_q}}$ to the $(q-1)^{th}$ extensions, $\eta^i_{j_1 j_2 \dots j_{q-1}}$, and adds the term

$$- \frac{D\xi_\nu}{Dx_{j_q}} u^i{}_{,j_1 j_2 \dots j_{q-1}\nu}$$

where

$$j_\alpha = 1, 2, \dots, n , \quad \text{for} \quad \alpha = 1, 2, \dots, q \quad .$$

<u>Reduction of the Number of Variables of a System of Partial Differential Equations Through Invariance</u>

Without loss of generality, we shall restrict ourselves to a system S of second order partial differential equations:

$$H_\alpha(x_j, u^i, u^i{}_{,j}, u^i{}_{,jk}) = 0 \quad ,$$
$$\alpha = 1, 2, \dots, r \quad , \tag{2.2-9}$$

on which boundary conditions

$$B_\beta(u^i{}_{,j}, u^i, x_j) = 0$$

are prescribed on curves

$$\omega_\beta(x_1, x_2, \dots, x_n) = 0 \ , \quad \beta = 1, 2, \dots, s \ . \tag{2.2-10}$$

Say $u^i = \Theta^i(x_1, x_2, \dots, x_n)$ is a solution to system S. Consider the system S^* obtained from S by having u^i replaced by v^i, x_j by $x^*_j = X^*_j(x_1,x_2,\dots,x_n, \Theta^1(x_1,x_2,\dots,x_n), \dots, \Theta^m(x_1,x_2,\dots,x_n); \varepsilon)$.

<u>Definition 2.2-1</u> .

The group of transformations (2.2-2) leaves S <u>invariant</u> iff

$v^i = U^{i*}(x_1,x_2,\dots,x_n,\ \Theta^1(x_1,x_2,\dots,x_n),\ \dots,\ \Theta^m(x_1,x_2,\dots,x_n);\ \varepsilon)$ satisfies S^* whenever $u^i = \Theta^i(x_1,x_2,\dots,x_n)$ is a solution to S.

In other words, if $u^i = \Theta^i(x_1,x_2,\dots,x_n)$ solves S and u^i is replaced by $\Theta^i(x_1,x_2,\dots,x_n)$ in (2.2-2), then S is invariant if:

(i) the system of partial differential equations (2.2-9) is invariant, i.e.,

$$H_\alpha(x_j^*,\ u^{*i},\ u^{*i}{}_{,j},\ u^{*i}{}_{,jk}) = 0\ ,\qquad \alpha = 1,\ 2,\ \dots,\ r \tag{2.2-11}$$

and

(ii) the boundary conditions and boundary curves also take on the same form in the new variables, i.e.,

$$B_\beta(u^{*i}{}_{,j},\ u^{*i},\ x_j^*) = 0$$

on

$$\omega_\beta(x_1^*,\ x_2^*,\ \dots,\ x_n^*) = 0\quad ,\ \text{for each}\quad \beta = 1,\ 2,\ \dots,\ s\ . \tag{2.2-12}$$

Introducing the infinitesimal elements of the extended transformations we see that (cf. §2.1) (2.2-11) is satisfied iff

$$\xi_j \frac{\partial H_\alpha}{\partial x_j} + \eta^i \frac{\partial H_\alpha}{\partial u^i} + \eta^i_j \frac{\partial H_\alpha}{\partial u^i{}_{,j}} + \eta^i_{jk} \frac{\partial H_\alpha}{\partial u^i{}_{,jk}} = 0$$

$$\text{for}\quad \alpha = 1,\ 2,\ \dots,\ r \tag{2.2-13}$$

i.e. each H_α, considered as a function of the $[\frac{m}{2}(n+2)(n+1)+n]$ variables $\{x_j,\ u^i,\ u^i{}_{,j},\ u^i{}_{,jk}\}$, is an invariant of the extended group of transformations of the group (2.2-2).

In order to show that invariance under a one-parameter group of transformations reduces the number of variables by one, it is more convenient to work with canonical variables. Recall that in §1.5 it was shown that a one-parameter group of transformations of two variables is equivalent to a translation group in terms of canonical variables where one of the variables is held fixed (i.e. is an invariant) and the other is translated. It is left as an exercise for the reader to show that this result can be extended to $k \geq 3$ variables, i.e. canonical variables can be found where $k - 1$ of the canonical variables are invariants of the group and the other one is translated.

We now introduce the canonical variables for (2.2-2). We label the new independent variables $\{y_1, y_2, \ldots, y_n\}$ and the new dependent variables $\{w_1, w_2, \ldots, w_m\}$.[1] The transformation from (x, u) coordinates to (y,w) coordinates can be expressed in the form:

$$x_j = X_j(y_1,y_2,\ldots,y_n, w^1,w^2,\ldots,w^m) ,$$
$$u^i = f^i(y_1,y_2,\ldots,y_n, w^1,w^2,\ldots,w^n) , \tag{2.2-14}$$

for some functions X_j, f^i, $j = 1, 2, \ldots, n$, $i = 1, 2, \ldots, m$.

Subsituting (2.2-14) in (2.2-9,10) we obtain a new system S' :

$$I_\alpha(y_j, w^i, \frac{\partial w^i}{\partial y_j}, \frac{\partial^2 w^i}{\partial y_j \partial y_k}) = 0 , \alpha = 1,2,\ldots,r \tag{2.2-15}$$

on which boundary conditions

$$C_\beta(\frac{\partial w^i}{\partial y_j}, w^i, y_j) = 0$$

are prescribed on curves

$$\nu_\beta(y_1, y_2, \ldots, y_n) = 0 ,$$
$$\beta = 1, 2, \ldots, s . \tag{2.2-16}$$

---

(1) Deciding whether a given canonical variable is dependent or independent is not necessarily obvious and the choice is often not unique (eg. the hodograph transformation interchanges dependent and independent variables).

The transformation from system S to system S' will be discussed later in this section.

Without loss of generality, in terms of the canonical variables, the group (2.2-2) leaving S invariant, becomes the group

$$\begin{aligned} y_1^* &= y_1 + \varepsilon \\ y_j^* &= y_j, \quad j = 2, \ldots, n \\ w^{*i} &= w^i, \quad i = 1, 2, \ldots, m \end{aligned} \tag{2.2-17}$$

leaving S' invariant. Thus I_α , the boundary conditions, and the boundary curves do not depend explicitly on y_1 .

We will now show that the solutions to S' also have this property.

Let

$$u^i = U^i(x_1, x_2, \ldots, x_n) \ , \quad i = 1,2,\ldots,m \tag{2.2-18}$$

be a solution to S . Then the corresponding solution

$$w^i = W^i(y_1, y_2, \ldots, y_n) \ , \quad i = 1,2,\ldots,m \tag{2.2-19}$$

to system S' is obtained by making use of (2.2-14) and solving the algebraic equations:

$$\begin{aligned} &U^i(X_1(y,w), X_2(y,w), \ldots, X_n(y,w)) = \\ &f^i(y_1,y_2,\ldots,y_n, w^1,w^2,\ldots,w^m) \ , \quad i = 1,2,\ldots,m \ . \end{aligned} \tag{2.2-20}$$

Now we assume that S , and hence S' , has a unique solution. If S' has a solution of the form (2.2-19), then in terms of the starred variables, we have two candidates for the solution w^{*i} in the starred S' system:

$$\text{(i)} \quad w^{*i} = w^i = W^i(y_1, y_2, \ldots, y_n) \ , \quad i = 1,2,\ldots,m \tag{2.2-21}$$

because of the invariance property, and,

$$\text{(ii)} \quad w^{*i} = W^i(y_1^*, y_2^*, \ldots, y_n^*) \ , \quad i = 1,2,\ldots,m \tag{2.2-22}$$

because if (2.2-19) is a solution to S' , then (2.2-22) is a solution to the starred S' system.

But the solution of w^{*i} is unique. Hence, for each $i = 1,2,\ldots,m$

$$\begin{aligned} W^i(y_1, y_2, \ldots, y_n) &= W^i(y_1^*, y_2^*, \ldots, y_n^*) \\ &= W^i(y_1 + \varepsilon, y_2, \ldots, y_n) \end{aligned} \tag{2.2-23}$$

which proves that the solutions W^i to S' do not depend on y_1 . Thus the number of variables has been reduced by one provided S is equivalent to S' . We now show how this transformation from S to S' is accomplished and under what conditions it is possible.

The Transformation from S to S'

In order to pass from S to S' we need to show how and under what conditions the derivative terms transform. Referring back to (2.2-20) we recall that any solution $u^i = U^i(x_1, x_2, \ldots, x_n)$ to (2.2-9,10) constructed from invariance under the group of transformations (2.2-2) can be written in terms of the canonical variables either in the form

$$u^i = U^i(X_1(y,w), X_2(y,w), \ldots, X_n(y,w)) \ ,$$

$i = 1, 2, \ldots, m$, or in the form

$$u^i = f^i(y_1, y_2, \ldots, y_n, w^1, w^2, \ldots, w^m) \ ,$$

$i = 1, 2, \ldots, m$. Hence for each i , the second representation of the solution leads to

$$\frac{Du^i}{Dy_j} = \frac{\partial f^i}{\partial y_j} + \frac{\partial f^i}{\partial w^\mu}\frac{\partial w^\mu}{\partial y_j} \tag{2.2-24}$$

whereas the first representation of the solution leads to

$$\frac{Du^i}{Dy_j} = \frac{\partial u^i}{\partial x_\nu}\left[\frac{\partial X_\nu}{\partial y_j} + \frac{\partial X_\nu}{\partial w^\mu}\frac{\partial w^\mu}{\partial y_j}\right] = \frac{\partial u^i}{\partial x_\nu}\frac{DX_\nu}{Dy_j} \quad . \tag{2.2-25}$$

Hence solving (2.2-24,25) for $\frac{\partial u^i}{\partial x_j}$ we get

$$\frac{\partial u^i}{\partial x_j} = \frac{\dfrac{D(X_1,X_2,\ldots,X_{j-1},f^i,X_{j+1},\ldots,X_n)}{D(y_1,y_2,\ldots\ldots\ldots\ldots\ldots\ldots,\ldots,y_n)}}{\dfrac{D(X_1,X_2,\ldots\ldots\ldots\ldots\ldots\ldots,\ldots,X_n)}{D(y_1,y_2,\ldots\ldots\ldots\ldots\ldots\ldots,\ldots,y_n)}} \tag{2.2-26}$$

and thus we require that the Jacobian determinant

$$\frac{D(X_1,X_2,\ldots,X_n)}{D(y_1,y_2,\ldots,y_n)} = \begin{vmatrix} \frac{DX_1}{Dy_1} & \cdots & \frac{DX_1}{Dy_n} \\ \vdots & & \vdots \\ \frac{DX_n}{Dy_1} & & \frac{DX_n}{Dy_n} \end{vmatrix}$$

$$= \det \left\| \frac{DX_i}{Dy_j} \right\| \neq 0 \tag{2.2-27}$$

i.e., the transformation from x to y is locally one-to-one. Moreover, the generated solution to S must satisfy (2.2-27) for S to be equivalent to S' .

The transformation of second order derivatives is found in an analogous fashion, by calculating

$$\frac{D^2u^i}{Dy_jDy_k} = \frac{\partial u^i}{\partial x_\nu}\frac{D^2X_\nu}{Dy_jDy_k} + \frac{\partial^2 u^i}{\partial x_\nu \partial x_\omega}\frac{DX_\nu}{Dy_j}\frac{DX_\omega}{Dy_k} \tag{2.2-28}$$

or alternatively,

$$\frac{D^2u^i}{Dy_jDy_k} = \frac{\partial^2 f^i}{\partial y_j \partial y_k} + \frac{\partial^2 f^i}{\partial y_j \partial w^\nu}\frac{\partial w^\nu}{\partial y_k} + \frac{\partial^2 f^i}{\partial y_k \partial w^\nu}\frac{\partial w^\nu}{\partial y_j}$$
$$+ \frac{\partial^2 f^i}{\partial w^\lambda \partial w^\mu}\frac{\partial w^\lambda}{\partial y_j}\frac{\partial w^\mu}{\partial y_k} + \frac{\partial f^i}{\partial w^\mu}\frac{\partial^2 w^\mu}{\partial y_j \partial y_k} \quad . \tag{2.2-29}$$

One can show that [2]:

$$\det \left\| \frac{DX_\nu}{Dy_j} \frac{DX_\omega}{Dy_k} \right\| = \det \left\| \frac{DX_i}{Dy_j} \otimes \frac{DX_i}{Dy_j} \right\|$$

$$= \left[\det \left\| \frac{DX_i}{Dy_j} \right\| \right]^{2n} \qquad (2.2\text{-}30)$$

Thus, for example

$$\frac{\partial^2 u^i}{\partial x_1^2} = \frac{\begin{vmatrix} \frac{D^2 f^i}{Dy_1^2} - \frac{\partial u^i}{\partial x_\nu}\frac{D^2 X_\nu}{Dy_1^2} & \frac{DX_1}{Dy_1}\frac{DX_2}{Dy_1} & \cdots & \frac{DX_n}{Dy_1}\frac{DX_n}{Dy_1} \\ \frac{D^2 f^i}{Dy_1 Dy_2} - \frac{\partial u^i}{\partial x_\nu}\frac{D^2 X_\nu}{Dy_1 Dy_2} & \frac{DX_1}{Dy_1}\frac{DX_2}{Dy_2} & \cdots & \frac{DX_n}{Dy_1}\frac{DX_n}{Dy_2} \\ \vdots & \vdots & & \vdots \\ \frac{D^2 f^i}{Dy_n^2} - \frac{\partial u^i}{\partial x_\nu}\frac{D^2 X_\nu}{Dy_n^2} & \frac{DX_1}{Dy_n}\frac{DX_2}{Dy_n} & \cdots & \frac{DX_n}{Dy_n}\frac{DX_n}{Dy_n} \end{vmatrix}}{\left[\mathrm{Det} \left\| \frac{DX_i}{Dy_j} \right\| \right]^{2n}} \qquad (2.2\text{-}31)$$

In practice canonical variables are not used when applying group methods to reduce the number of variables appearing in a system of partial differential equations. We work directly with the given variables as illustrated in §2.1. A discussion of invariance under a multi-parameter group is left to a later chapter. In the next three chapters we discuss dimensional analysis and its relationship to invariance of partial differential equations under a group of stretching transformations. Our prime example will be the heat equation.

---

(2) W. H. Greub, Multilinear Algebra, Springer-Verlag, 1967, p. 26.

2.3 Fundamental Solution of the Heat Equation; Dimensional Analysis

Before proceeding any further, we digress and show how in certain special cases dimensional analysis can be used to obtain the same results as are attainable from group theory.

The fundamental solution of the heat equation gives the temperature distribution in an infinite medium due to the concentrated addition of a (unit) source of heat. The solution has various interpretations and applications which are discussed later.

Consider first one-dimensional heat conduction in an infinite bar with constant properties. Initially, at time $(t = 0-)$, the bar is at uniform temperature $u = u_o = 0$ (say, without loss of generality). The mathematical representation of this problem is

$$\begin{aligned} &\rho c \frac{\partial u}{\partial t} - k \frac{\partial^2 u}{\partial x^2} = Q\delta(x)\delta(t) , \qquad -\infty < x < \infty, \quad t \geq 0- \\ &u(x,0-) = 0 \\ &u(\pm \infty, t) \to 0 . \end{aligned} \tag{2.3-1}$$

In (2.3-1) the source is at the origin but if at any other location x_o we need merely replace x by $x - x_o$. This is physically evident and can be made to follow formally from the translation invariance. If the temperature is restricted in a physically reasonable way (e.g. bounded below for $Q > 0$) then the problem specified by (2.3-1) has a unique solution.

The fundamental principle of dimensional analysis enables the form of the solution to be written down in this case. The fundamental principle of dimensional analysis states: every problem must be able to be expressed in dimensionless variables or alternatively all equalities must involve only dimensionally consistent quantities. The fundamental principle is a statement of invariance of all physical problems with respect to choice of units of measurement. The quantities entering Problem (2.3-1) all have physical dimensions:

Physical Constants[(1)]

ρ = density $\{\rho\} = \frac{K}{M^3}$

c = specific heat $\{c\} = \frac{cal}{deg \cdot K}$

k = thermal conductivity $\{k\} = \frac{cal}{SM^2} \cdot \frac{M}{deg} = \frac{cal}{degSM}$

Q = heat added per cross-section area $\{Q\} = \frac{cal}{M^2}$

Variables

u = temperature $\{u\} = deg$

x = space coordinate $\{x\} = M$

t = time coordinate $\{t\} = S$

According to the general principle the independent variables (x,t) should be able to be expressed in dimensionless variables but an examination of the physical constants shows that no combination of them can provide a physical constant with dimensions either of length (M) or time (S) which could be used to make x or t dimensionless[(2)].

The problem has no characteristic length or time scale. Note however, the thermal diffusivity κ :

---

(1) For this problem the following units will be used: mass = Kilograms = K , length = meters = M , time = seconds = S , heat = calories = cal. , temperature = degrees abs. = deg. { } = dimensions of .

(2) If a characteristic temperature u_c existed, then (x,t) could be made dimensionless. We are implicitly using the fact that the initial temperature u is not characteristic and only $(u - u_o)$ can appear.

$$\kappa = \frac{k}{\rho c} \qquad \{\kappa\} = \frac{M^2}{S} \; .$$

Hence a dimensionless variable can only be formed from a suitable combination of (x,t), namely,

$$z = \frac{x}{2\sqrt{\kappa t}} \; . \tag{2.3-2}$$

In order to make the dependent variable u dimensionless a quantity with the dimensions of degrees must be formed with the help of the physical constants. The quantity, $Q/\rho c$ has dimensions

$$\left\{\frac{Q}{\rho c}\right\} = M \cdot \text{deg} \; .$$

Since no characteristic temperature exists no quantity with dimensions of purely (deg.) can be found. However by using the length x, or $\sqrt{\kappa t}$ a dimensionless combination

$$\frac{u\sqrt{\kappa t}}{Q/\rho c} \tag{2.3-3}$$

can be formed. Thus, the functional form of the solution is completely defined

$$u(x,t;\rho c,k,Q) = \frac{Q}{\rho c\sqrt{\kappa t}} \, f(z) \tag{2.3-4}$$

where $f(z)$ is a dimensionless function. $f(z)$ must satisfy an ordinary differential equation.

The analogous problem in two or three dimensions is different only in the dimensions of the heat addition term. Thus,

$$\text{two-dimensions} \qquad \{Q\} = \frac{\text{heat added}}{\text{length}} = \frac{\text{cal}}{M} \; ; \; u = \frac{Q}{\rho c\kappa t} \, f(z_2) \tag{2.3-5}$$

$$\text{three-dimensions} \qquad \{Q\} = \text{heat added} = \text{cal} \; ; \; u = \frac{Q}{\rho c(\kappa t)^{3/2}} \, f(z_3) \tag{2.3-6}$$

where

$$z_2 = \frac{\sqrt{x^2+y^2}}{2\sqrt{\kappa t}} = \frac{r}{2\sqrt{\kappa t}} \quad , \qquad r = \text{cylindrical radius} \quad ,$$

$$z_3 = \frac{\sqrt{x^2+y^2+z^2}}{2\sqrt{\kappa t}} = \frac{r}{2\sqrt{\kappa t}} \quad , \qquad r = \text{spherical radius} \quad .$$

The connection between dimensional analysis and invariance of partial differential equations under stretching transformations will be discussed in §2.5.

2.4 Fundamental Solutions of Heat Equation; Global Affinity

In this section we treat the problem of the previous section by studying its invariance under global transformations, in particular stretching transformations. The method is thus analogous to that used in §1.1 in introducing the ideas for ordinary differential equations. In effect we see what can be done without the use of infinitesimal transformations.

Our problem is (2.3-1) and we let the temperature field be

$$u = \Theta(x,t) \tag{2.4-1}$$

so that (cf. 2.3-1)

$$\begin{aligned} \frac{\partial\Theta}{\partial t} - \kappa \frac{\partial^2\Theta}{\partial x^2} &= \frac{Q}{\rho c}\,\delta(x)\,\delta(t) \\ \Theta(x,0-) &= 0 \\ \Theta(\pm\,\infty,t) &\to 0 \end{aligned} \tag{2.4-2}$$

i.e., $u = \Theta(x,t)$ is the solution to (2.3-1).

We consider the general stretching transformation of the (u,x,t) space:

$$\begin{aligned} u^* &= \gamma u \\ x^* &= \alpha x \\ t^* &= \beta t \end{aligned} \tag{2.4-3}$$

with parameters (γ,α,β). If $\gamma(\beta)$, $\alpha(\beta)$ are somehow determined then (2.4-3) is a one-parameter group of transformations with the identity element $\gamma = \alpha = \beta = 1$.[1] Corresponding to (2.4-1) we have a new surface defined by

$$u^* = \Theta^*(x^*,t^*) \quad . \tag{2.4-4}$$

We ask how the original solution surface transforms:

$$\frac{\partial\Theta}{\partial t} = \frac{\beta}{\gamma}\frac{\partial\Theta^*}{\partial t^*} , \quad \frac{\partial\Theta}{\partial x} = \frac{\alpha}{\gamma}\frac{\partial\Theta^*}{\partial x^*} , \quad \frac{\partial^2\Theta}{\partial x^2} = \frac{\alpha^2}{\gamma}\frac{\partial^2\Theta^*}{\partial x^{*2}} \quad .$$

Since $\Theta(x,t)$ is defined by (2.4-2) we have

$$\frac{\beta}{\gamma}\frac{\partial\Theta^*}{\partial t^*} - \kappa\frac{\alpha^2}{\gamma}\frac{\partial^2\Theta^*}{\partial x^{*2}} = \frac{Q}{\rho c}\,\delta\left(\frac{x^*}{\alpha}\right)\delta\left(\frac{t^*}{\beta}\right) \quad . \tag{2.4-5}$$

For invariance it is necessary that both the operators on the left and the right hand side of (2.4-5) agree with those in (2.4-2), multiplied by a common factor. Thus, for invariance

$$\alpha^2 = \beta \; ; \qquad \alpha = \sqrt{\beta} \quad . \tag{2.4-6}$$

Further, it follows from the integral definition of the δ-fns:

$$\int \delta(x)dx = 1 \; , \quad \int \delta(ax)dx = \frac{1}{a}\int \delta(ax)d(ax)$$

that

$$\delta(ax) = \frac{1}{a}\,\delta(x) \quad . \tag{2.4-7}$$

Thus (2.4-5) becomes

$$\frac{\beta}{\gamma}\left[\frac{\partial\Theta^*}{\partial t^*} - \kappa\frac{\partial^2\Theta^*}{\partial x^{*2}}\right] = \beta^{3/2}\,\frac{Q}{\rho c}\,\delta(x^*)\delta(t^*) \tag{2.4-8}$$

For invariance of the equation then

$$\gamma = \frac{1}{\sqrt{\beta}} \quad . \tag{2.4-9}$$

---

(1) If $\beta = e^{\varepsilon}$, then $\varepsilon = 0$ corresponds to the identity.

Our one parameter family of transformations of the (u,x,t) space to itself is thus (cf. 2.4-3)

$$u^* = \frac{1}{\sqrt{\beta}} u$$
$$x^* = \sqrt{\beta}\, x \tag{2.4-10}$$
$$t^* = \beta t$$

(2.4-10) is the group of transformations leaving invariant Problem (2.3-1). Under this transformation the initial and boundary conditions attached to (2.3-1) are also invariant. Thus, for Θ^* we have

$$\frac{\partial \Theta^*}{\partial t^*} - \kappa \frac{\partial^2 \Theta^*}{\partial x^{*2}} = \frac{Q}{\rho c}\, \delta(x^*)\,\delta(t^*)$$
$$\Theta^*(x^*, 0-) = 0 \tag{2.4-11}$$
$$\Theta^*(\pm\infty, t^*) \to 0 \quad .$$

Now, due to the uniqueness, Θ must be the <u>same</u> function of (x^*,t^*) as Θ^* is of (x,t). That is

$$\Theta^*(x,t) = \Theta(x^*,t^*) \tag{2.4-12}$$

As a consequence of the transformation (2.4-10) and the invariance condition (2.4-12) we thus obtain a functional equation which must be satisfied by the solution:

$$u^* = \Theta(x^*,t^*) = \frac{1}{\sqrt{\beta}} u = \frac{1}{\sqrt{\beta}} \Theta(x,t)$$

or

$$\Theta(\sqrt{\beta}x, \beta t) = \frac{1}{\sqrt{\beta}} \Theta(x,t) \quad . \tag{2.4-13}$$

(2.4-13) holds for all values of β. From this functional relation the functional form that the solutions $\Theta(x,t)$ must have can be deduced in various ways. For example we can say that the factor

$1/\sqrt{\beta}$ is a scaling like $1/\sqrt{t}$ from (2.4-10) and a coordinate like $x/\sqrt{t}$ is invariant so that necessarily

$$\Theta(x,t) = \frac{1}{\sqrt{t}} f\left(\frac{x}{\sqrt{t}}\right)^{(2)} \quad . \tag{2.4-14}$$

Evidently (2.4-13) is satisfied and within trivial changes (2.4-14) is unique. Alternatively consider $\partial/\partial\beta$ of (2.4-13) near the identity $(\beta = 1)$ (that is, study the infinitesimal form of (2.4-13)).

$$\frac{1}{2\sqrt{\beta}}\, x \frac{\partial\Theta}{\partial x}(\sqrt{\beta}x,\beta t) + t \frac{\partial\Theta}{\partial t}(\sqrt{\beta}x,\beta t) = -\frac{1}{2\beta^{3/2}}\Theta(x,t) \tag{2.4-15}$$

As $\beta \to 1$, we see that $\Theta(x,t)$ must also satisfy a first-order p.d.e.

$$\frac{1}{2}\, x \frac{\partial\Theta}{\partial x}(x,t) + t \frac{\partial\Theta}{\partial t}(x,t) = -\frac{1}{2}\Theta(x,t) \quad . \tag{2.4-16}$$

This enables the form of the solution to be found since the general solution of (2.4-16) involves an arbitrary function. The characteristic equations associated with (2.4-16) are

$$\frac{dx}{\frac{1}{2}x} = \frac{dt}{t} = -\frac{d\Theta}{\frac{1}{2}\Theta} \quad . \tag{2.4-17}$$

The integral of the first two is

$$\zeta = \frac{x}{2\sqrt{t}} \tag{2.4-18}$$

and the integral of the second two is

$$\log \Theta = -\frac{1}{2}\log t + \log F(\zeta) \quad . \tag{2.4-19}$$

Thus, the general solution of (2.4-16) has the form

(2) Alternatively $\frac{x}{\sqrt{t}}$ and $\sqrt{t}\Theta$ are two functionally independent invariants of (2.4-10). The solution form (2.4-14) is obtained by setting one invariant as an arbitrary function of the other.

$$\Theta(x,t) = \frac{1}{\sqrt{t}} F(\zeta) \quad , \quad \zeta = \frac{x}{2\sqrt{t}} \quad . \tag{2.4-20}$$

This agrees exactly with the form (2.3-4) derived by dimensional analysis; introducing the dimensional quantities let

$$T = \Theta(x,t) = \frac{Q}{\rho c\sqrt{\kappa t}} f(z) \quad , \quad z = \frac{x}{2\sqrt{\kappa t}} \quad . \tag{2.4-21}$$

We now proceed to the solution by deriving the ordinary differential equation for f . We have

$$\frac{\partial T}{\partial t} = \frac{Q}{\rho c} \frac{1}{\sqrt{\kappa}} \left[\frac{1}{\sqrt{t}} \frac{df}{dz} \left(\frac{-z}{2t} \right) - \frac{1}{2t^{3/2}} f(z) \right]$$

$$= \frac{-Q}{\rho c\sqrt{\kappa}} \frac{1}{2t^{3/2}} \left[z \frac{df}{dz} + f \right]$$

$$\frac{\partial^2 T}{\partial x^2} = \frac{Q}{\rho c \; 4(\kappa t)^{3/2}} \frac{d^2 f}{dz^2}$$

so that for $t \geq 0+$ the partial differential equation of (2.4-2) becomes

$$\frac{d^2 f}{dz^2} + 2z \frac{df}{dz} + 2f = 0 \quad . \tag{2.4-22}$$

The boundary and initial conditions of (2.4-2) are now (cf. Figure 2.4-1)

$$f(\pm\infty) \to 0 \quad . \tag{2.4-23}$$

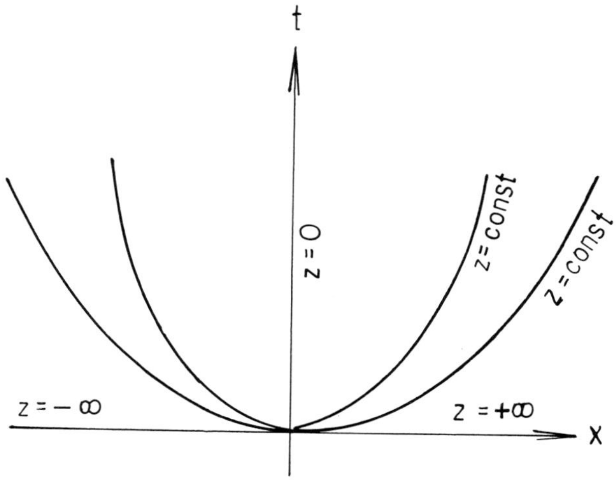

Figure 2.4-1

The condition (2.4-23) is not sufficient to define the solution; account must be taken of the fact that an amount of heat Q has been introduced to the medium. The law of conservation of total heat, or $\int_{0-}^{t} dt \int_{-\infty}^{\infty} dx$ of the heat equation is

$$\rho c \int_{-\infty}^{\infty} \Theta(x,t)\,dx = Q \quad . \tag{2.4-24}$$

Using the similarity form (2.4-21), (2.4-24) becomes

$$\int_{-\infty}^{\infty} \frac{1}{\sqrt{\kappa t}} f(z)\,dx = 1$$

or

$$\int_{-\infty}^{\infty} f(z)\,dz = \frac{1}{2} \quad . \tag{2.4-25}$$

(2.4-25) is the extra condition needed to define the solution uniquely. Now the solutions to (2.4-22) can be expressed in terms of Hermite functions and of course the solution to the problem here is well known.

We adopt another course here in order to illustrate the application of group theory to the ordinary differential equation (2.4-22). This will give some idea of what to expect in more complicated cases. The equation (2.4-22) is linear and so admits the group

$$\begin{aligned} f_1 &= \alpha f \\ z_1 &= z \quad . \end{aligned} \tag{2.4-26}$$

Thus, invariant coordinates are

$$v = \frac{1}{f}\frac{df}{dz} \tag{2.4-27}$$

and z and the equation is reduced to one of Riccati type. (cf example (ii) page 112 of §1.17). We have

$$\frac{dv}{dz} = \frac{1}{f}\frac{d^2 f}{dz^2} - \frac{1}{f^2}\left(\frac{df}{dz}\right)^2$$

$$= \frac{-2z\frac{df}{dz} - 2f}{f} - v^2$$

or

$$\frac{dv}{dz} = -v^2 - 2zv - 2 \quad . \tag{2.4-28}$$

Along an integral curve of (2.4-28) the quadrature for f can be expressed((2.4-27))

$$\frac{df}{f} = v\,dz \quad . \tag{2.4-29}$$

A sketch of all the possible paths of (2.4-28) is easily drawn with the help of the zero slope isoclines

$$v^2 + 2zv + 2 = 0 \quad . \tag{2.4-30}$$

See Fig. (2.4-2).

f is always positive for solutions of interest. The arrows indicate the direction of f increasing. Since the solution must have the same qualitative behavior as $z \to \pm\infty$ it is clear that only the exceptional path which separates the two classes of paths has a

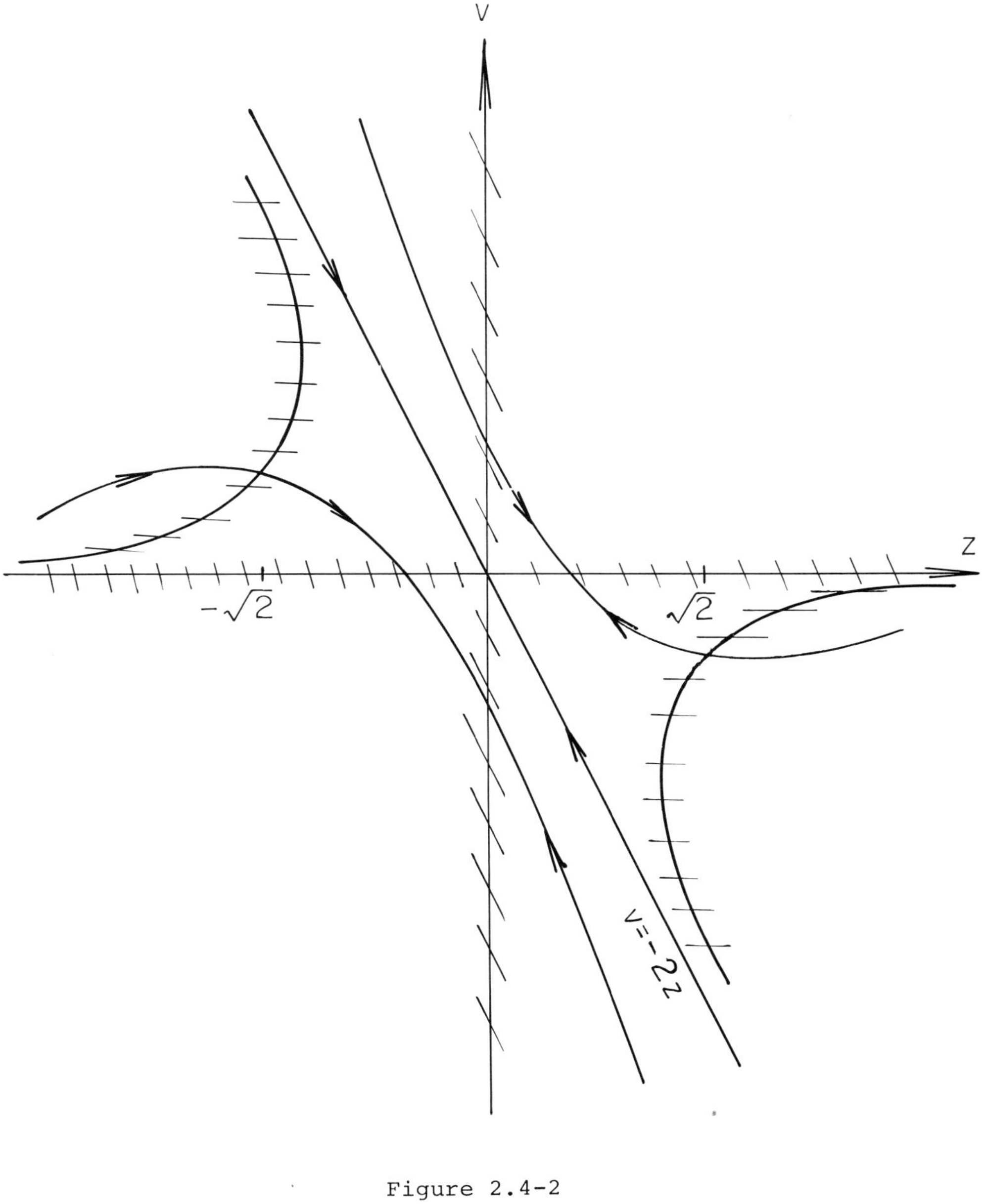

Figure 2.4-2

chance to be the correct path for the given boundary conditions. As $z \to \pm \infty$ the quadratic terms cancel if $v = -2z$ and this defines the exceptional path near ∞. However,

$$v = -2z \tag{2.4-31}$$

is an exact integral of (2.4-28) and gives a straight line through the origin. Hence on this path

$$\frac{df}{f} = -2z\,dz \quad . \tag{2.4-32}$$

The integral is

$$f = Ce^{-z^2} . \tag{2.4-33}$$

The remaining constant of integration C is found from the integral condition (2.4-25)

$$C \int_{-\infty}^{\infty} e^{-z^2} dz = \frac{1}{2}$$

or

$$C = \frac{1}{2\sqrt{\pi}} \quad . \tag{2.4-34}$$

Hence the complete solution (2.4-21) is

$$\begin{aligned} u &= \Theta(x,t) \\ &= \frac{Q/\rho c}{2\sqrt{\pi\kappa t}}\, e^{-\frac{x^2}{4\kappa t}} \quad , \end{aligned} \tag{2.4-35}$$

the well-known result.

Problem 2.4-1. Using the fact that one solution (2.4-31) of the Riccati equation (2.4-28) is known, construct the general solution

$$f = e^{-z^2}\left[C + D \int_0^z e^{\zeta^2}\, d\zeta\right]$$

which corresponds to all the other paths in Fig. 2.4-2 if $D \neq 0$ (cf pp. 95 and 96 of §1.15).

It is worth noting that the same argument applies in the cylindrical and spherical cases and the solution can be carried out in the same way. Corresponding to (2.4-35) we have

$$u_2 = \theta_2(r,t)$$

$$= \frac{Q/\rho c}{4\pi\kappa t} e^{\frac{-r^2}{4\kappa t}} \quad , \quad r^2 = x^2 + y^2 \quad ,$$

$$u_3 = \theta_3(r,t) \tag{2.4-36}$$

$$= \frac{Q/\rho c}{(4\pi\kappa t)^{3/2}} e^{\frac{-r^2}{4\kappa t}} \quad , \quad r^2 = x^2 + y^2 + z^2 \quad .$$

Some remarks can be made about other interpretations of the fundamental solution.

(1) As a far field - if heat is added to a one-dimensional bar over finite length and time, then after a sufficiently long time the solution should approach the fundamental solution with the same total heat added. Formally, by superposition,

$$u = \theta(x,t) = \int_A\int \frac{q(\xi,\tau)/\rho c}{2\sqrt{\pi\kappa(t-\tau)}} e^{\frac{-(x-\xi)^2}{4\kappa(t-\tau)}} d\xi \, d\tau \tag{2.4-37}$$

where A represents the finite domain in which heat is added and q the density of heat addition. (See Fig. 2.4-3)

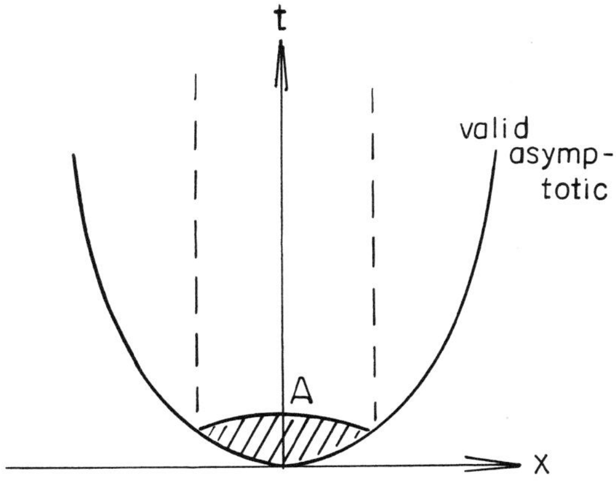

Figure 2.4-3

It is clear that for $t >> \tau$, $|x| >> \xi$:

$$\theta \to \frac{1/\rho c}{2\sqrt{\pi\kappa t}} \left(\iint_A q(\xi,\tau)\, d\xi\, d\tau \right) e^{-\frac{x^2}{4\kappa t}} . \tag{2.4-38}$$

A similar idea is often applied to the far field of non-linear problems although the idea of superposition is of course no longer valid. The proper interpretation is given in terms of perturbation theory in which, eventually, the far field appears in suitable coordinates (x,t) in which the size of the disturbance area A tends toward zero. Thus a point singularity appears and a corresponding similarity solution. In these problems however it is not always clear what constants (corresponding to total heat, for example) characterize the solution.

(2) As a local solution - let the heat source of (2.4-1) be located at $x = x_o$ in a finite bar $0 \le x \le \ell$ whose ends are kept at zero temperature

$$\theta(0,t) = \theta(\ell,t) = 0 . \tag{2.4-39}$$

If the heat source at the origin in the infinite medium has the solution (2.4-35)

$$u = S(x,t) = \frac{Q/\rho c}{2\sqrt{\pi\kappa t}} e^{-\frac{x^2}{4\kappa t}} \tag{2.4-40}$$

then the problem with the boundary condition (2.4-39) is solved by reflection. (See Fig. 2.4-4).

$$u = S(x - x_o,t) - S(x + x_o,t) + S(x - 2\ell - x_o,t)$$
$$- S(x - 2\ell + x_o,t) + S(x + 2\ell - x_o,t) + \ldots$$

or

$$u = \sum_{n=-\infty}^{\infty} \{ S(x - x_o - 2n\ell,t) - S(x + x_o + 2n\ell,t) \} . \tag{2.4-41}$$

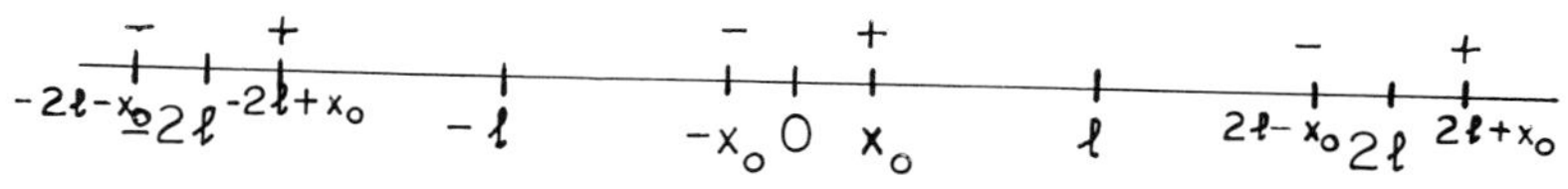

Figure 2.4-4

It is clear that the first term representing the source at $x = x_0$ is a good approximation near $x = x_0$ up to such times as the effect of the boundaries is felt. The solution is thus only locally similar. After a long time the representation (2.4-41) is not a good one, since all terms in the sum are important - a representation in Fourier series over the interval is better. This problem does have an overall length ℓ so that overall similarity certainly cannot exist. When such ideas are tried in non-linear problems interaction of the non-linear solutions must be taken into account.

Problem 2.4-2

Apply the reasoning of this section to obtain the similarity form for the boundary value problem in $t > 0$, $x > 0$:

$$\frac{\partial \theta}{\partial t} - \frac{\partial^2 \theta}{\partial x^2} = 0 \quad , \qquad \theta(0,t) = t^{\alpha} \quad , \qquad \theta(x,0) = 0 \quad .$$

2.5 The Relationship Between the Use of Dimensional Analysis and Stretching Groups to Reduce the Number of Variables of a Partial Differential Equation

In this section we will give a short discussion of dimensional analysis and its applicability to reducing the number of variables of a partial differential equation. Dimensional analysis concerns the invariance of formulae under the group of scalings of the fundamental units of the variables and (physical) constants entering the formulae.[(1),(2),(3)] Its applicability to partial differential equations may lead to a reduction of the number of independent variables.[(3)]

Solutions obtained through dimensional analysis are called self-similar solutions of the first type.[4] The group method described in §2.1 involves transformations of the variables with all (physical) constants held fixed. It will be shown (Theorem 2.5-2) that if the number of variables entering a partial differential equation can be reduced by $\rho > 0$ through dimensional analysis (where the variables and constants are both stretched) then the number of variables can also be reduced by ρ from invariance of the p.d.e. under a ρ-parameter group of stretchings of the variables with the constants held fixed. By example it will be shown that the converse of this theorem is false, i.e., there exist solutions to problems obtained from invariance of a partial differential equation[5] under a group of stretching transformations which cannot be obtained through dimensional analysis.[6] Hence the application of group methods to construct solutions to partial differential equations is a generalization of the concept of dimensional analysis.

Given any (physical) problem there will be defined m fundamental dimensional units (eg. mass, length, and time) $L_1, L_2, \ldots, L_m$ in which each quantity entering the problem will be measured. Quantities are of two types - variables and (physical) constants. Say u is

(1) P. W. Bridgman, Dimensional Analysis, Second Edition, Yale University Press, 1931.

(2) L. I. Sedov, Similarity and Dimensional Methods, Moscow, Sixth Edition (in Russian); English translation, Fourth Edition, Academic Press, 1959.

(3) G. Birkhoff, Hydrodynamics, Second Edition, Princeton University Press, 1960.

(4) G. I. Barenblatt and Ya. B. Zel'dovich, Self-similar solutions as intermediate asymptotics, Ann. Rev. of F. M., 1972.

(5) In this paragraph the words "partial differential equation" are understood to include all associated boundary conditions of the problem in question.

(6) Such solutions are called self-similar solutions of the second type (cf. Barenblatt and Zel'dovich, ibid.)

the unknown quantity entering a problem[7] and that the other n quantities are labelled $W_1, W_2, \ldots, W_n$. Our objective is to find u as a function of these n quantities, i.e., to find the function f such that

$$u = f(W_1, W_2, \ldots, W_n) \quad . \tag{2.5-1}$$

Dimensional analysis restricts the choice of f as seen in the following theorem, the so-called Buckingham Pi-Theorem. This theorem forms the basis for modelling.

Theorem 2.5-1 (Buckingham Pi-Theorem)

We make the following three assumptions:

(i) If Z is a measurable quantity connected with our given problem, then $\{Z\} = L_1^{\alpha_1} L_2^{\alpha_2} \ldots L_m^{\alpha_m}$, for some real numbers $\alpha_1, \alpha_2, \ldots, \alpha_m$, where $\{\ \}$ denotes dimension of. The vector $\alpha = [\alpha_1, \alpha_2, \ldots, \alpha_m]$ is the dimension vector of Z.

(ii) (2.5-1) is invariant under an arbitrary scaling (stretching) of any fundamental dimensional unit; i.e. the formula holds for any set of dimensional units. (eg. MKS, CGS, or Br. Eng. system)

(iii) u is a differentiable function of $W_1, W_2, \ldots, W_n$.

Under these assumptions we prove that:

(a) The formula (2.5-1) for finding u can be expressed in terms of dimensionless quantities.[8]

(b) The number of dimensionless quantities (called π-variables) is $k + 1 = n + 1 - q$, where q is the

---

(7) In order to simplify the discussion we assume the problem at hand has only one unknown.

(8) Z is a dimensionless quantity iff $\{Z\} = 1$.

rank of the $(m)\times(n+1)$ dimension matrix $\mathcal{B}$ formed from the dimensions (exponents) of the $n+1$ quantities $u, W_1, W_2, \ldots, W_n$.

(c) The formula (2.5-1) for finding u can be simplified to

$$u = G(W_1,W_2,\ldots,W_n)F(\pi_1,\pi_2,\ldots,\pi_k) \qquad (2.5\text{-}2)$$

where

$$G = W_1^{\beta_1}W_2^{\beta_2}\ldots W_n^{\beta_n}$$

for some real numbers $\beta_1, \beta_2, \ldots, \beta_n$,

$$\pi_i = W_1^{\gamma_{i1}}W_2^{\gamma_{i2}}\ldots W_n^{\gamma_{in}}$$

for some real numbers $\gamma_{i1}, \gamma_{i2}, \ldots, \gamma_{in}$, $i = 1, 2, \ldots, k$, and F is an arbitrary function of $\pi_1, \pi_2, \ldots, \pi_k$. Moreover $\frac{u}{G}, \pi_1, \ldots, \pi_k$ are dimensionless quantities.

Proof:

By assumption (i)

$$\{u\} = L_1^{a_1}L_2^{a_2}\ldots L_m^{a_m}$$

and

$$\{W_i\} = L_1^{b_{i1}}L_2^{b_{i2}}\ldots L_m^{b_{im}} \qquad (2.5\text{-}3)$$

for some real numbers a_j, b_{ij}, $i = 1, 2, \ldots, n$, $j = 1, 2, \ldots, m$.

Corresponding to the quantities $u, W_1, \ldots, W_n$ we define the dimension matrix

$$\mathcal{B} = \begin{bmatrix} b_{11} & b_{12} & \cdots & b_{1m} \\ \vdots & \vdots & & \vdots \\ b_{n1} & b_{n2} & \cdots & b_{nm} \\ a_1 & a_2 & \cdots & a_m \end{bmatrix}, \qquad (2.5\text{-}4)$$

We now consider the invariance of (2.5-1) under arbitrary scalings of the fundamental units by taking each unit in turn (assumption (ii)). We scale L_1 by letting $L_1^* = e^{\varepsilon_1} L_1$, where ε_1 is an arbitrary real number. Under this transformation the quantities transform as follows:

$$\begin{aligned} W_i^* &= e^{\varepsilon_1 b_{i1}} W_i, \quad i = 1,2,\ldots,n \\ u^* &= e^{\varepsilon_1 a_1} u, \end{aligned} \qquad (2.5\text{-}5)$$

(2.5-5) defines a one-parameter (ε_1) Lie group of stretching transformations of the $n+1$ variables and (physical) constants where $\varepsilon_1 = 0$ corresponds to the identity element. Note that this Lie group of transformations on the variables and constants is induced by the one-parameter stretching group (scalings) acting on the dimensional unit L_1.

Say (2.5-1) happened to be the formula relating $u, W_1, W_2, \ldots, W_n$. Then by assumption (ii):

$$u^* = f(W_1^*, W_2^*, \ldots, W_m^*) \quad . \qquad (2.5\text{-}6)$$

Expanding (2.5-6) about $\varepsilon_1 = 0$ and equating the $0(\varepsilon_1)$ terms, we find that f satisfies the <u>invariant surface condition</u> (cf. §2.1)

$$a_1 f(W_1,\ldots,W_n) = \sum_{i=1}^{n} b_{i1} W_i \frac{\partial f}{\partial W_i}(W_1,\ldots,W_n) \quad . \qquad (2.5\text{-}7)$$

The following arguments can now be made from (2.5-7).

If $b_{i1} = 0$, $i = 1, 2, \ldots, n$, $a_1 \neq 0$, then $u \equiv 0$.
If $b_{i1} = 0$, $i = 1, 2, \ldots, n$, $a_1 = 0$, then L_1 is not a fundamental dimensional unit for the given quantities.
If $b_{i1} \neq 0$ for some i, without loss of generality we assume that $b_{11} \neq 0$. The characteristic equations corresponding to (2.5-7) are:

$$\frac{du}{a_1 u} = \frac{dW_1}{b_{11}W_1} = \frac{dW_2}{b_{21}W_2} = \cdots = \frac{dW_n}{b_{n1}W_n} \quad . \qquad (2.5\text{-}8)$$

Integrating the last $n - 1$ equalities of (2.5-8) we find that

$$X_{i-1} = W_i W_1^{-b_{i1}/b_{11}} \quad , \quad i = 2, 3, \ldots, n \quad , \qquad (2.5\text{-}9)$$

are invariants of (2.5-5). Integrating the first equality of (2.5-8) we find that

$$u_1 = uW_1^{-a_1/b_{11}} \qquad (2.5\text{-}10)$$

is also an invariant of (2.5-5). Thus

$$u = W_1^{\frac{a_1}{b_{11}}} f_1(X_1, X_2, \ldots, X_{n-1}) \qquad (2.5\text{-}11)$$

where f_1 is some arbitrary function of $X_1, X_2, \ldots, X_{n-1}$. Hence the number of quantities in the arbitrary function on which u depends has been reduced by one. Moreover $u_1, X_1, X_2, \ldots, X_{n-1}$ are invariants under arbitrary scalings of L_1.

Proceeding, we now scale L_2 by letting $L_2^* = e^{\varepsilon_2} L_2$ for arbitrary ε_2. Then we induce the following Lie group of stretching transformations acting on $u_1, X_1, X_2, \ldots, X_{n-1}$:

$$X_{i-1}^* = e^{\frac{\varepsilon_2}{b_{11}}(b_{11}b_{i2} - b_{12}b_{i1})} X_{i-1} \quad , \quad i = 2,3,\ldots,n \quad ,$$

$$u_1^* = e^{\frac{\varepsilon_2}{b_{11}}(b_{11}a_2 - b_{12}a_1)} u_1 \quad . \qquad (2.5\text{-}12)$$

If $b_{11}b_{i2} - b_{12}b_{i1} \neq 0$ for some i, without loss of generality we can assume that $b_{11}b_{22} - b_{12}b_{21} \neq 0$. Letting $c_i = (b_{11}b_{i2}-b_{12}b_{i1})/(b_{11}b_{22}-b_{12}b_{21})$, $i = 3, 4, \ldots, n$, we find that

$$Y_{i-2} = X_{i-1}X_1^{-c_i} \quad , \quad i = 3, 4, \ldots, n \quad , \tag{2.5-13}$$

and

$$u_2 = u_1 X_1^{-(b_{11}a_2-b_{12}a_1)/(b_{11}b_{22}-b_{12}b_{21})} \tag{2.5-14}$$

are invariants of the Lie group of stretchings (2.5-12).

Hence

$$u_2 = f_2(Y_1, Y_2, \ldots, Y_{n-2})$$

and thus

$$u = W_1^{a_1/b_{11}} X_1^{(b_{11}a_2-b_{12}a_1)/(b_{11}b_{22}-b_{12}b_{21})} f_2(Y_1, Y_2, \ldots, Y_{n-2}) \tag{2.5-15}$$

where f_2 is some arbitrary function of $Y_1, Y_2, \ldots, Y_{n-2}$.

$\{u_2, Y_1, Y_2, \ldots, Y_{n-2}\}$ are invariants under arbitrary scalings of L_1 and L_2 . By induction it is easy to show that the invariance of (2.5-1) under arbitrary scalings of $L_1, L_2, \ldots, L_m$, implies that u has the form (2.5-2) with $\{\frac{u}{G}\} = \{\pi_i\} = 1$, $i = 1, 2, \ldots, k$. The number k of π products can now be studied and we can show that $k = n - q$, where $q = r(\mathcal{B})$ is the rank of the dimension matrix $\mathcal{B}$ (2.5-4).

Let the vector $x = [x_1, x_2, \ldots, x_{n+1}]$ where $x_i \in R$, $i = 1, 2, \ldots, n+1$.

We first note that $W_1^{x_1} W_2^{x_2} \ldots W_n^{x_n} u^{x_{n+1}}$ is invariant under arbitrary scaling of L_j , $j = 1, 2, \ldots, m$, iff x satisfies the equation

$$x\mathcal{B} = 0 \quad . \tag{2.5-16}$$

The number of linearly independent solutions x of (2.5-16) is equal to $n + 1 - r(\mathcal{B})$.

Hence, since the dimensionless quantities $\frac{u}{G}$, $\pi_1, \ldots, \pi_k$ are all of the form $W^{d_1} W^{d_2} \ldots W^{d_n} u^d$ (from the inductive proof) for some real numbers $d_1, d_2, \ldots, d_n, d,$ we see that the number of functionally independent invariants (dimensionless quantities) under the m-parameter Lie group of scalings of $L_1, L_2, \ldots, L_m$ is equal to $k + 1 = n + 1 - q$ where $q = r(\mathcal{B})$. This completes the proof of Theorem 2.5-1.

We note that $r(\mathcal{B})$ can be replaced by $r(B)$ where

$$B = \begin{bmatrix} b_{11} & b_{12} & \cdots & b_{1m} \\ \vdots & & & \vdots \\ b_{n1} & b_{n2} & \cdots & b_{nm} \end{bmatrix} , \tag{2.5-17}$$

since if $r(B) \neq r(\mathcal{B})$ then $u \equiv 0$. (In this case none of the dimensionless quantities depend on u) Without loss of generality we assume that $r(B) = r(\mathcal{B})$.

In reducing the class of formulae for the functional form of u from (2.5-1) to (2.5-2) we have made no use of any (differential) equations relating the $n + 1$ quantities $u, W_1, W_2, \ldots, W_n$. In the case where we know a set of governing equations relating these $n + 1$ quantities, the conclusions of Theorem 2.5-1 imply that the equations can be made dimensionless by using the "π-variables". Say that in the case when we are dealing with a system of partial differential equations $u, W_1, W_2, \ldots, W_\ell$ are the <u>variables</u> and $W_{\ell+1}, W_{\ell+2}, \ldots, W_n$ are the (physical) <u>constants</u> appearing in the equations. We assume that our system of equations contains only one dependent variable u .

We set $B = \begin{bmatrix} B_1 \\ B_2 \end{bmatrix}$

$$\text{where } B_1 = \begin{bmatrix} b_1 & b_{12} & \cdots & b_{1m} \\ \vdots & & & \vdots \\ b_{\ell 1} & b_{\ell 2} & \cdots & b_{\ell m} \end{bmatrix} \tag{2.5-18}$$

and

$$B_2 = \begin{bmatrix} b_{\ell+1,1} & b_{\ell+1,2} & \cdots & b_{\ell+1,m} \\ \cdot & & & \\ \cdot & & & \\ \cdot & & & \\ b_{n1} & b_{n2} & \cdots\cdots & b_{nm} \end{bmatrix} . \qquad (2.5\text{-}19)$$

B_1 and B_2 are respectively the dimension matrices of the independent variables and (physical) constants entering the problem at hand.

If $r(B_2) = r(B)$, then through dimensional analysis the number of variables appearing in the differential equations is not reduced, i.e., after dimensional analysis the number of variables appearing in the equations is still $\ell + 1$, although the number of constants has been reduced by $r(B)$.[9] Thus, in effect, if $r(B_2) = r(B)$, the differential equations have not been simplified in the sense of reducing the number of variables, although through the dimensional analysis non-dimensional constants appear. These are useful for example in the construction of perturbation expansions.

If $r(B) = r(B_2) + \rho$, $\rho > 0$, then in terms of the dimensionless π-quantities the number of variables appearing in the given equations can be reduced by ρ whereas the number of constants can be reduced by $r(B_2)$.

In §2.1, §2.2 and §2.4 we have discussed the invariance of partial differential equations under group transformations of the variables without regard to the dimensions of the variables. Naturally this leads one to consider the following question: If the number of variables appearing in a system of partial differential equations (including associated boundary conditions) can be reduced by ρ

---

(9) The quantity $W_1^{d_1}, W_2^{d_2}, \ldots, W_n^{d_n} u^d$ is called a <u>constant</u> iff $d = d_1 = d_2 = \ldots = d_\ell = 0$; otherwise this quantity is called a <u>variable</u>.

through dimensional analysis, can this reduction be accomplished through invariance of the system under a ρ parameter group of stretchings acting on the variables only, i.e., is the method of dimensional analysis for reducing the number of variables in differential equations included in the method of group invariance of differential equations? Theorem 2.5-2 answers this question in the affirmative.

Theorem 2.5-2

If the number of variables appearing in a system of partial differential equations can be reduced by ρ through dimensional analysis (which involves transforming both variables and constants), then the number of variables can be reduced by ρ through invariance of the partial differential equations under a ρ-parameter group of stretching transformations applied to its variables only.

Proof:

As before let

$$B = \begin{bmatrix} B_1 \\ B_2 \end{bmatrix} .$$

Let $r = r(B_2)$, and let $q = r + \rho = r(B)$. Then through dimensional analysis the number of variables is reduced by ρ.

$$\text{Let} \quad \varepsilon = \begin{pmatrix} \varepsilon_1 \\ \varepsilon_2 \\ \cdot \\ \cdot \\ \cdot \\ \varepsilon_m \end{pmatrix} \neq \begin{pmatrix} 0 \\ 0 \\ \cdot \\ \cdot \\ \cdot \\ 0 \end{pmatrix}$$

Let $L_i^* = e^{\varepsilon_i} L_i$, $i = 1,2,\ldots,m$.

Then

$$W_i^* = e^{\sum_{s=1}^{m} b_{is}\varepsilon_s} W_i$$

$$= e^{(B\varepsilon)_i} W_i \quad , \quad i = 1,2,\ldots,n \quad ,$$

where $(B\varepsilon)_i$ is the ith component of the vector $B\varepsilon$, and

$$u^* = e^{\sum_{s=1}^{m} a_s\varepsilon_s} u \tag{2.5-20}$$

The induced group of transformations (2.5-20) leaves invariant our system of equations for <u>any</u> values of $\varepsilon_1, \varepsilon_2, \ldots, \varepsilon_m$ because of invariance under dimensional scalings.

$$W_i^* = W_i \; , \; i = \ell + 1, \ell + 2, \ldots, n$$

$$\text{iff} \qquad B_2\varepsilon = 0 \quad . \tag{2.5-21}$$

We now introduce a few definitions and appropriate notation. Let A be a linear transformation on the vector space V . Then the <u>null space of A</u> , $V_{A_N} = \{\varepsilon \in V\colon \ A\varepsilon = 0\}$ and the <u>range space of A</u> , $V_{A_R} = \{x;\ x = A\varepsilon$ for some $\varepsilon \in V\}$. <u>dim V</u> is the dimension of the vector space V , $\dim V = \dim V_{A_R} + \dim V_{A_N}$.

Consider B , B_1 and B_2 as defined by (2.5-17,18,19) . $\dim(R^m)_{B_N}$ = number of linearly independent solutions of the set of equations $B\varepsilon = 0$. $\dim(R^m)_{B_{2N}} = m - r$, $\dim(R^m)_{B_N} = m - r - \rho$.

Note that

$$\dim(R^m)_{B_{2N}} = \dim((R^m)_{B_{2N}})_{B_{1N}}$$

$$+ \dim((R^m)_{B_{2N}})_{B_{1R}}$$

$$= \dim(R^m)_{B_N} + \dim((R^m)_{B_{2N}})_{B_{1R}}$$

$$\text{since} \quad ((R^m)_{B_{2N}})_{B_{1N}} = (R^m)_{B_N} \quad .$$

$$\text{Hence} \quad \dim((R^m)_{B_{2N}})_{B_{1R}} = \rho \quad .$$

But $((R^m)_{B_{2N}})_{B_{1R}}$ contains all vectors corresponding to invariance under stretchings applied to the variables only. Hence the system of partial differential equations is invariant under a ρ-parameter group of stretching transformations applied to the variables only.

The following example will show that the converse of this theorem is false, i.e., reduction of the number of variables from invariance under a stretching group does not necessarily mean that dimensional analysis works.

The Rayleigh Flow Problem

As an example we consider the Rayleigh flow problem.[10] Say an

---

[10] Schlichting, H., Boundary Layer Theory, McGraw-Hill, New York, 1955, p. 64.

infinite flat plate is immersed in an incompressible fluid at rest. The plate is instantaneously accelerated so that it moves parallel to itself with velocity $U(t)$. We find the velocity distribution of the fluid.

Let u be the fluid velocity in the direction of $U(t)$ (x-direction). Let v and w be the fluid velocities respectively in the directions normal to the plate (y-direction) and tangential to the plate (z-direction). Then from symmetry considerations it is easy to see that $v = w = 0$ and that the Navier-Stokes equations reduce to the viscous diffusion equation

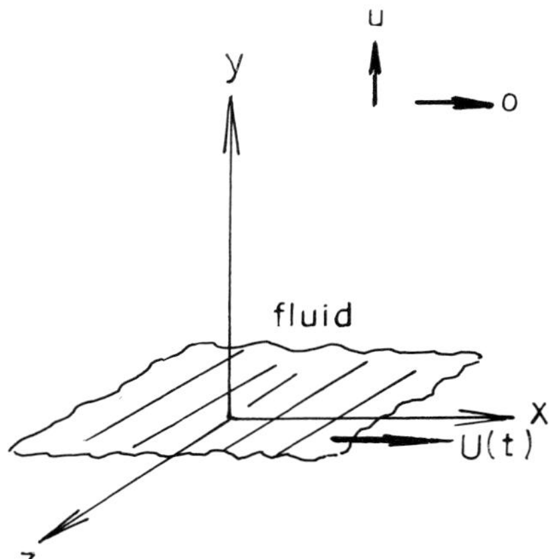

Figure 2.5-1

$$\rho \frac{\partial u}{\partial t} = \mu \frac{\partial^2 u}{\partial y^2} \quad . \tag{2.5-22}$$

Here ρ = fluid density and μ = coefficient of viscosity. For simplicity we use ν = kinematic viscosity = $\frac{\mu}{\rho}$. The boundary conditions are

$$u(y,0) = 0 \ , \quad y > 0$$

$$u(0,t) = U(t) \ , \quad t > 0 \tag{2.5-23}$$

$$u(y \to \infty, t) = 0 \quad .$$

We will consider the case $U(t)$ = constant = U from the points of view of dimensional analysis, dimensional analysis with "ingenuity" and invariance under a group of stretchings applied to the variables only. The case $U(t) \neq$ constant will be considered in an exercise of §2.8.

(i) Dimensional Analysis

The fundamental units are $L_1 = L$ = length and $L_2 = T$ = time

$\{u\} = LT^{-1}$ $\{U\} = LT^{-1}$

$\{t\} = T$ $\{\nu\} = L^2T^{-1}$

$\{y\} = L$

We set $W_1 = t$, $W_2 = y$, $W_3 = \nu$, $W_4 = U$.

The dimension matrices are:

$$\mathcal{B} = \begin{pmatrix} 0 & 1 \\ 1 & 0 \\ 2 & -1 \\ 1 & -1 \\ 1 & -1 \end{pmatrix}$$

$$B = \begin{pmatrix} 0 & 1 \\ 1 & 0 \\ 2 & -1 \\ 1 & -1 \end{pmatrix}$$

$$B_1 = \begin{pmatrix} 0 & 1 \\ 1 & 0 \end{pmatrix}$$

$$B_2 = \begin{pmatrix} 2 & -1 \\ 1 & -1 \end{pmatrix}$$

$r(B_2) = r(B) = 2$. Hence dimensional analysis will not lead to a reduction in the number of variables appearing in the partial differential equation. We now consider this problem from the point of view of dimensional analysis with a little added "insight".

(ii) Dimensional Analysis with "ingenuity"

As is easily seen if we let $u_1 = \frac{u}{U}$, then (2.5-22,23) become:

$$\frac{\partial u_1}{\partial t} = \nu \frac{\partial^2 u_1}{\partial y^2} \tag{2.5-24}$$

$$\begin{aligned} u_1(y,0) &= 0 , \quad y > 0 \\ u_1(0,t) &= 1 , \quad t > 0 \\ u_1(y \to \infty, t) &= 0 . \end{aligned} \tag{2.5-25}$$

Now applying dimensional analysis we see that

$$\{u_1\} = 1 , \text{ i.e. } u_1 \text{ is dimensionless}$$

$$\{t\} = T$$

$$\{y\} = L$$

$$\{\nu\} = L^2 T^{-1} .$$

We set $W_1 = t$, $W_2 = y$, $W_3 = \nu$ (W_4 has been eliminated).

$$\mathcal{B} = \begin{pmatrix} 0 & 1 \\ 1 & 0 \\ 2 & -1 \\ 0 & 0 \end{pmatrix}$$

$$B = \begin{pmatrix} 0 & 1 \\ 1 & 0 \\ 2 & -1 \end{pmatrix}$$

$$B_2 = \begin{pmatrix} 2 & -1 \end{pmatrix}$$

$$r(B_2) = 1 , \qquad r(B) = 2 .$$

$$\rho = r(B) - r(B_2) = 1 .$$

Hence using dimensional analysis after "appropriately" transforming (2.5-22,23) we are able to reduce the number of variables by one.

Applying dimensional analysis to (2.5-24,25) it is easy to see that the dimensionless quantities are

$$z = \frac{y}{\sqrt{\nu t}} \tag{2.5-26}$$

and

$$\frac{u}{U} = u_1 = F(z) \tag{2.5-27}$$

Substituting (2.5-27) into (2.5-24) we reduce this partial differential equation to an ordinary differential equation with dependent variable $F(z)$:

$$\frac{d^2F}{dz^2} + \frac{1}{2} z \frac{dF}{dz} = 0 \tag{2.5-28}$$

with boundary conditions

$$F(\infty) = 0 \quad , \qquad F(0) = 1 \quad . \tag{2.5-29}$$

Integrating (2.5-28) and using the boundary conditions (2.5-29), one can show that

$$F(z) = \operatorname{erfc} \frac{z}{2} = 1 - \frac{2}{\sqrt{\pi}} \int_0^{z/2} e^{-\zeta^2} d\zeta \quad . \tag{2.5-30}$$

(iii) Method of invariance under stretching group.

Say the set of stretching transformations applied to the variables only

$$u^* = e^{\varepsilon_1} u$$
$$t^* = e^{\varepsilon_2} t \tag{2.5-31}$$
$$y^* = e^{\varepsilon_3} y$$

leaves invariant (2.5-22,23). We find the relationship between ε_1, ε_2 and ε_3 such that these equations are invariant.

Invariance of (2.5-22)

$$\Rightarrow \qquad \varepsilon_2 = 2\varepsilon_3 \quad .$$

Invariance of (2.5-23)

$$\Rightarrow \qquad \varepsilon_1 = 0 \quad .$$

Hence the one-parameter (ε) group

$$\begin{aligned} u^* &= u \\ t^* &= e^{2\varepsilon} t \\ y^* &= e^{\varepsilon} y \end{aligned} \tag{2.5-32}$$

leaves invariant (2.5-22,23). Thus the number of variables can be reduced by one. (This shows that the converse of Theorem 2.5-2 is false.)

The invariants of (2.5-32) are u and $\frac{y}{\sqrt{t}}$. The similarity variable is

$$\omega = \frac{y}{\sqrt{t}} \quad . \tag{2.5-33}$$

Setting

$$u = f(\omega) \tag{2.5-34}$$

and substituting this expression into (2.5-22) we find that $f(\omega)$ satisfies the ordinary differential equation

$$\nu \frac{d^2 f}{d\omega^2} + \frac{\omega}{2} \frac{df}{d\omega} = 0 \tag{2.5-35}$$

with boundary conditions

$$\begin{aligned} f(\infty) &= 0 \quad , \\ f(0) &= U \quad . \end{aligned} \tag{2.5-36}$$

Thus

$$f(\omega) = U \operatorname{erfc} \frac{\omega}{2\sqrt{\nu}} \quad . \tag{2.5-37}$$

<u>Problem 2.5-1</u> (Boundary layer on a wedge)

Consider the problem of a boundary-layer flow on a semi-infinite wedge at zero angle of attack [11],[12]. The governing partial differential equations are

$$u \frac{\partial u}{\partial x} + v \frac{\partial u}{\partial y} - U(x) \frac{dU}{dx} = \nu \frac{\partial^2 u}{\partial y^2}$$
$$\frac{\partial u}{\partial x} + \frac{\partial v}{\partial y} = 0 \qquad (2.5\text{-}38)$$

with boundary conditions

$$u(x, 0) = v(x, 0) = 0$$
$$\lim_{y \to +\infty} u(x, y) = U(x) \quad . \qquad (2.5\text{-}39)$$

For flow past a wedge

$$U(x) = Ax^{\ell} \quad . \qquad (2.5\text{-}40)$$

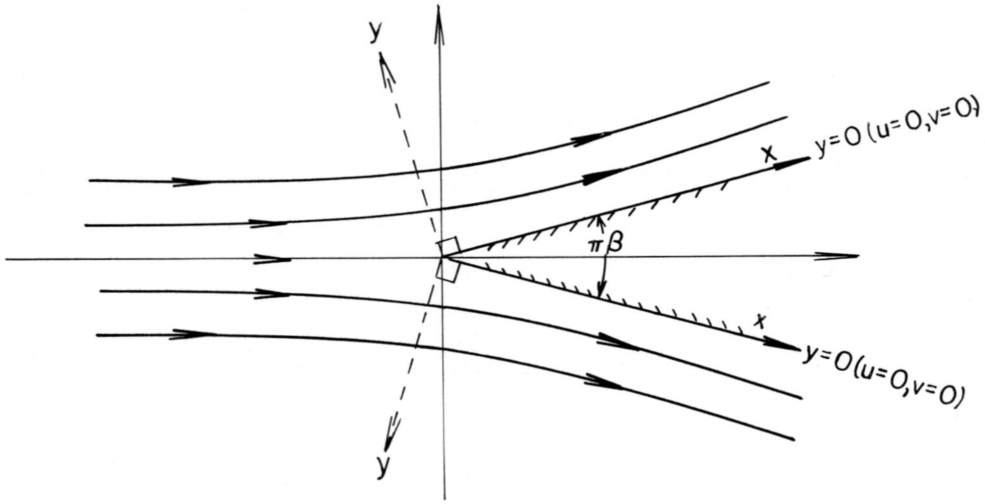

Figure 2.5-2

(11) P. A. Lagerstrom, High Speed Aerodynamics and Jet Propulsion, Vol. IV: Laminar Flows and Transition to Turbulence, Princeton University Press.

(12) Birkhoff, ibid (3).

x is the distance from the leading edge on the wedge surface (tangential coordinate), y is the distance from the wedge surface (normal coordinate), u is the x-component of velocity, v is the y-component of velocity, ν is the kinematic viscosity, A is a constant, and the constant $\ell = \frac{\beta}{2-\beta}$ where $\pi\beta$ is the opening angle of the semi-infinite wedge (see Figure 2.5-2).

Introduce the streamfunction $\psi(x, y)$:

$$u = -\frac{\partial\psi}{\partial y}, \quad v = \frac{\partial\psi}{\partial x},$$

so that the continuity equation of (2.5-38) is satisfied. Thus the system of equations (2.5-38,39) has dependent variable ψ, independent variables x and y, physical dimensional constants A and ν and a dimensionless constant ℓ.

(i) Show that if $\ell \neq 0$ (i.e. $\beta \neq 0$) then dimensional analysis is not fruitful in reducing the number of variables in (2.5-38).

(ii) For $\ell = 0$ use dimensional analysis to reduce (2.5-38,39) to a 3rd order ordinary differential equation (the Blasius equation (1.18-54)) and find the corresponding boundary conditions.

(iii) For arbitrary ℓ show that (2.5-38,39) is invariant under a one-parameter group of stretchings of the variables. Obtain the corresponding ordinary differential equation and boundary conditions. Choose your similarity coordinates so that the resulting o.d.e.[13] becomes the Blasius equation in the case $\ell = 0$.

§2.6 Use of Group Invariance to Obtain New Solutions from Given Solutions

Group invariance was used in §2.1 to §2.5 to reduce the number

---

(13) This ordinary differential equation is called the Falkner-Skan equation.

of variables of a partial differential equation and in Part 1 to reduce the order of an ordinary differential equation. In both cases the aim is the construction of solutions without knowledge of any solutions. Before proceeding in the remaining sections to apply group methods to various partial differential equations we now discuss how group invariance of differential equations may be used to obtain new solutions from given solutions[(1)]. Without loss of generality we restrict ourselves to considering second order differential equations with one dependent and at most two independent variables.

A very simple example of the type of thing we discuss here is the following: As shown in (2.5-37), a solution $u(y,t)$ of the heat equation derived from invariance under a stretching group in (y,t) is

$$u(y,t) = U \operatorname{erfc}\left(\frac{y}{2\sqrt{\nu t}}\right) .$$

A different group leaving the heat equation invariant is translation in y (or t). That is, $(y \to y - y_o,\ t \to t)$ leaves the equation the same. Therefore a new solution is

$$u(y,t) = U \operatorname{erfc}\left(\frac{y-y_o}{2\sqrt{\nu t}}\right) .$$

Note that no new solution is obtained if the stretching group is used on the original solution.

Consider (cf. §2.1) the partial differential equation S

$$H(u_{xx}, u_{xt}, u_{tt}, u_x, u_t, u, x, t) = 0 \qquad (2.6\text{-}1)$$

defined on domain D.

We consider a group of transformations $G = \{g\}$ (not necessarily a Lie group) where for each $g \in G$ the variables are transformed as

---

(1) In this section boundary conditions are absent.

follows:

$$\left.\begin{aligned} u^* &= U_g(x,\ t,\ u) \\ x^* &= X_g(x,\ t) \\ t^* &= T_g(x,\ t) \end{aligned}\right\}^{(2)} \tag{2.6-2}$$

and $D \to D_g$. We introduce the shorthand notations:

$$\mathrm{x} = (x,\ t)$$

$$\mathrm{x}^* = (x^*,\ t^*)$$

and

$$\mathrm{x}^* = g\mathrm{x} \quad \text{denotes} \quad \begin{aligned} x^* &= X_g(x,\ t) \\ t^* &= T_g(x,\ t) \end{aligned}$$

For any $g \in G$, there exists $g^{-1} \in G$ (the inverse element) such that

$$\mathrm{x} = g^{-1}\mathrm{x}^* \quad . \tag{2.6-3}$$

Say $u = \theta(\mathrm{x})$ is a solution to S on domain D . Consider the partial differential equation S_g defined on domain D_g obtained from S by having the dependent variable u replaced by dependent variable v , and the independent variables x by $\mathrm{x}^* = g\mathrm{x}$.

Definition 2.6-1

The group G leaves S invariant iff for every $g \in G$

$$v = U_g(\mathrm{x},\ \theta(\mathrm{x})) \tag{2.6-4}$$

is a solution to S_g whenever $u = \theta(\mathrm{x})$ is a solution to S . Now we will assume that the group G leaves S invariant.

Since $u = \theta(\mathrm{x})$ is a solution to S , $v = \theta(\mathrm{x}^*)$ is a solution

---

(2) The results presented in this section may be extended to the case where X_g, T_g depend on u .

to S_g. Hence we have shown that starting with the solution $v = \theta(x^*)$ to S_g on D_g we can generate another solution

$$v = U_g(g^{-1}x^*, \theta(g^{-1}x^*)) \tag{2.6-5}$$

on D due to the invariance of S under G. Note that (2.6-5) is a function defined on D_g. But S_g is the same partial differential equation as S except for a relabelling of independent variables by x^* instead of x. Hence if $u = \theta(x)$ is a solution to S on D then for <u>any</u> $g \in G$

$$u = U_g(g^{-1}x, \theta(g^{-1}x)) \tag{2.6-6}$$

is a solution to S on the domain $D_g \cap D$. Thus invariance of (2.6-1) under a group G leads to new solutions from a known solution provided

$$U_g(g^{-1}x, \theta(g^{-1}x)) \neq \theta(x)$$

for every $g \in G$.

<u>Theorem 2.6-1</u>

If G defines a one-parameter (ε) Lie group of transformations and $\theta(x)$ is a solution to S constructed from invariance of S under G (cf. §2.1) then for every $g \in G$

$$U_g(g^{-1}x, \theta(g^{-1}x)) = \theta(x) \quad , \tag{2.6-7}$$

<u>Proof</u> :

If $u = \theta(x)$ is a solution to S constructed from invariance of S under G then for any $g \in G$, $\theta(x)$ satisfies (cf (2.1-5))

$$U_g(g^{-1}x^*, \theta(g^{-1}x^*)) = \theta(x^*) \tag{2.6-8}$$

for any $x^* \in D_g \cap D \neq \emptyset$ since (2.6-2) is now a Lie group. Hence by a suitable relabelling of x^* (2.6-7) follows for $x \in D \cap D_g$.

Use of infinitesimal operator to compute new solutions

In §1.5 the infinitesimal operator of a Lie group was introduced. Associated with the one-parameter (ε) Lie group of transformations

$$u^* = U^*(x,t,u;\ \varepsilon) = u + \varepsilon\eta(x,t,u) + O(\varepsilon^2)$$

$$x^* = X^*(x,t;\ \varepsilon) = x + \varepsilon\xi(x,t) + O(\varepsilon^2) \qquad (2.6\text{-}9)$$

$$t^* = T^*(x,t;\ \varepsilon) = t + \varepsilon\tau(x,t) + O(\varepsilon^2)$$

is the infinitesimal operator

$$\mathfrak{X} = \xi(x,t)\frac{\partial}{\partial x} + \tau(x,t)\frac{\partial}{\partial t} + \eta(x,t,u)\frac{\partial}{\partial u} \qquad (2.6\text{-}10)$$

In some neighbourhood of $\varepsilon = 0$

$$u^* = e^{\varepsilon\mathfrak{X}}u$$

$$= \left(\sum_{n=0}^{\infty}\frac{\varepsilon^n\mathfrak{X}^n}{n!}\right)u$$

$$= u + \varepsilon\eta(x,t,u) + \frac{\varepsilon^2}{2!}\left[\xi(x,t)\frac{\partial\eta}{\partial x} + \tau(x,t)\frac{\partial\eta}{\partial t} + \eta(x,t,u)\frac{\partial\eta}{\partial u}\right] + \ldots$$

$$x^* = e^{\varepsilon\mathfrak{X}}x$$

$$= x + \varepsilon\xi(x,t) + \frac{\varepsilon^2}{2!}\left[\xi\frac{\partial\xi}{\partial x} + \tau\frac{\partial\xi}{\partial t}\right] + \ldots$$

$$t^* = e^{\varepsilon\mathfrak{X}}t$$

$$= t + \varepsilon\tau(x,t) + \frac{\varepsilon^2}{2!}\left[\xi\frac{\partial\tau}{\partial x} + \tau\frac{\partial\tau}{\partial t}\right] + \ldots \qquad (2.6\text{-}11)$$

For any C^∞ function $\mathcal{F}(x,t,u)$,

$$\mathcal{F}(x^*,t^*,u^*) = e^{\varepsilon\mathfrak{X}}\mathcal{F}(x,t,u) \qquad (2.6\text{-}12)$$

In terms of the Lie group (2.6-9), if $u = \theta(x,t)$ is a solution to (2.6-1) then provided $\theta(x,t)$ is not constructed from invariance under (2.6-9) we generate a one-parameter (ε) family of solutions

$$u_\varepsilon = U(X^*(x,t;-\varepsilon),\ T^*(x,t;-\varepsilon),\ \theta(X^*(x,t;-\varepsilon),\ T^*(x,t;-\varepsilon));\ \varepsilon) \tag{2.6-13}$$

where $\varepsilon = 0 \leftrightarrow u_0 = \theta(x,t)$.

In terms of the infinitesimal operator $\mathfrak{X}$ (2.6-13) can be written in the form

$$\begin{aligned} u_\varepsilon &= U(e^{-\varepsilon\mathfrak{X}}x,\ e^{-\varepsilon\mathfrak{X}}t,\ \theta(e^{-\varepsilon\mathfrak{X}}x,\ e^{-\varepsilon\mathfrak{X}}t);\ \varepsilon) \\ &= e^{-\varepsilon\mathfrak{X}}U(x,t,\theta(x,t);\ \varepsilon) \qquad \text{by (2.6-12)} \\ &= e^{-\varepsilon\mathfrak{X}}\left[(e^{\varepsilon\mathfrak{X}}u)\Big|_{u=\theta(x,t)}\right] . \end{aligned} \tag{2.6-14}$$

From (2.6-11)

$$\mathscr{G}(x,t) = (e^{\varepsilon\mathfrak{X}}u)\Big|_{u=\theta(x,t)} = \theta(x,t) + \varepsilon\eta(x,t,\theta(x,t))$$

$$+ \frac{\varepsilon^2}{2!}\left[\xi(x,t)\frac{\partial\eta}{\partial x} + \tau(x,t)\frac{\partial\eta}{\partial t} + \eta(x,t,u)\frac{\partial\eta}{\partial u}\right]\Big|_{u=\theta(x,t)} + \cdots .$$

Hence

$$\begin{aligned} u_\varepsilon &= e^{-\varepsilon\mathfrak{X}}\mathscr{G}(x,t) \\ &= e^{-\varepsilon[\xi(x,t)\frac{\partial}{\partial x} + \tau(x,t)\frac{\partial}{\partial t}]}\mathscr{G}(x,t) \\ &= e^{-\varepsilon\mathfrak{X}}\theta(x,t) + \varepsilon e^{-\varepsilon\mathfrak{X}}\eta(x,t,\theta(x,t)) + \frac{\varepsilon^2}{2!}e^{-\varepsilon\mathfrak{X}}[\quad] + \cdots \\ &= \theta(x,t) + \varepsilon[\eta(x,t,\theta(x,t)) - \xi\theta_x - \tau\theta_t] \\ &\quad + \frac{\varepsilon^2}{2!}[\xi^2\theta_{xx} + \xi\xi_x\theta_x + \tau\xi_t\theta_x + 2\tau\xi\theta_{xt} + \xi\tau_x\theta_t + \\ &\qquad \tau^2\theta_{tt} + \tau\tau_t\theta_t + \eta\eta_\theta - \xi\eta_x - 2\xi\eta_\theta\theta_x - \tau\eta_t - 2\tau\eta_\theta\theta_t] + \cdots \\ &= \theta_0(x,t) + \varepsilon\theta_1(x,t) + \varepsilon^2\theta_2(x,t) + \cdots \end{aligned} \tag{2.6-15}$$

where $\theta_0(x,t) = \theta(x,t)$,

$$\theta_1(x,t) = \eta(x,t,\theta(x,t)) - \xi\theta_x - \tau\theta_t \;, \quad \text{etc.} \tag{2.6-16}$$

If $\theta(x,t)$ is a solution constructed from invariance under (2.6-9) then

$$\theta_i(x,t) = 0 \;, \quad i = 1, 2, \ldots$$

Note that $\theta_1(x,t) = 0$ corresponds to $\theta(x,t)$ satisfying the <u>invariant surface condition</u> (2.1-8).

<u>Special case - linear homogeneous equation</u>

Say (2.6-1) is a linear homogeneous partial differential equation, i.e., it is a p.d.e. of the form

$$\begin{aligned} Lu = {} & a(x,t)u_{xx} + b(x,t)u_{xt} + c(x,t)u_{tt} \\ & + d(x,t)u_x + e(x,t)u_t + k(x,t)u = 0 \end{aligned} \tag{2.6-17}$$

where L is the linear differential operator

$$L = a(x,t)\frac{\partial^2}{\partial x^2} + b(x,t)\frac{\partial^2}{\partial x\partial t} + c(x,t)\frac{\partial^2}{\partial t^2} + d(x,t)\frac{\partial}{\partial x} + e(x,t)\frac{\partial}{\partial t} + k(x,t) \tag{2.6-18}$$

<u>Theorem 2.6-2</u>

Say (2.6-17) is invariant under (2.6-9) and $\theta_o(x,t)$ is a solution of (2.6-17). Then for arbitrary $c_k \in C$, $k = 0, 1, 2, \ldots, n$,

$$u = \sum_{k=0}^{n} c_k\theta_k(x,t) \tag{2.6-19}$$

is a solution of (2.6-17).

<u>Proof</u>.

Invariance of (2.6-17) under (2.6-9) implies that (2.6-15, 16)

$$u_\varepsilon = e^{-\varepsilon \mathfrak{X}} \mathscr{G}(x,t)$$
$$= \sum_{j=0}^{\infty} \varepsilon^j \theta_j(x,t)$$

is a one-parameter family of solutions of (2.6-17) in some neighborhood of $\varepsilon = 0$.

$$Lu_\varepsilon = 0$$

iff

$$\sum_{j=0}^{\infty} \varepsilon^j L\Theta_j = 0$$

iff

$$L\Theta_j = 0 \ , \ j = 0, 1, 2, \ldots$$

Hence (2.6-19) is a solution of (2.6-17).

Problem 2.6-1

Consider the heat equation

$$u_{xx} - u_t = 0 \quad . \tag{2.6-20}$$

(i) By "inspectional analysis" find a four-parameter group of stretchings and translations leaving invariant (2.6-20).

(ii) Say $u = \theta(x,t)$ is a solution of (2.6-20). Find a four-parameter family of "new" solutions corresponding to invariance under the group of (i).

(iii) Apply the result of (ii) to the solution $u = \theta(x,t) = x^2-t$ and explain why only a three-parameter family of solutions is generated.

2.7 The General Similarity Solution of the Heat Equation

Derivation of the group of the heat equation

In this section we apply the formulation of §2.1 to the one-dimensional heat equation

$$u_{xx} - u_t = 0 \quad . \tag{2.7-1}$$

Lie[1] found the group of the heat equation although he did not proceed any further to show how group invariance leads to the construction of solutions to partial differential equations. Bluman[2] and Bluman and Cole[3] rederived the group of the heat equation and constructed the "general" similarity solution of the heat equation.

We first show how to derive the Lie group of transformations with infinitesimals (η, ξ, τ)

$$\begin{aligned} u^* &= U^*(x,t,u;\varepsilon) = u + \varepsilon\eta(x,t,u) + 0(\varepsilon^2) \\ x^* &= X^*(x,t,u;\varepsilon) = x + \varepsilon\xi(x,t,u) + 0(\varepsilon^2) \\ t^* &= T^*(x,t,u;\varepsilon) = t + \varepsilon\tau(x,t,u) + 0(\varepsilon^2) \end{aligned} \tag{2.7-2}$$

leaving invariant (2.7-1).

From (2.1-20, 21), we see that the invariant solution $u = \Theta(x,t)$ satisfies

$$u^*_{x^*x^*} - u^*_{t^*} = \Theta_{xx} - \Theta_t + \varepsilon(\eta_{xx} - \eta_t) + 0(\varepsilon^2)$$

where

(1) Sophus Lie, Über die Integration durch bestimmte Integral von einer Klasse linearer partieller Differentialgleichungen, Arch. for Math. Vol. VI, No. 3, Kristiana, 1881, p. 328.

(2) G. Bluman, Construction of Solutions to Partial Differential Equations by the Use of Transformation Groups, Ph.D. thesis, California Institute of Technology, 1967, Chap. II.

(3) George W. Bluman and Julian D. Cole, The General Similarity Solution of the Heat Equation, Journal of Math. and Mech., Vol. 18, No. 11, May, 1969, pp. 1025-1042.

$$\eta_{xx} - \eta_t = \left(\frac{\partial^2\eta}{\partial x^2} - \frac{\partial\eta}{\partial t}\right) + \left(2\frac{\partial^2\eta}{\partial x\partial u} - \frac{\partial^2\xi}{\partial x^2} + \frac{\partial\xi}{\partial t}\right)\Theta_x$$

$$+ \left(\frac{\partial\tau}{\partial t} - \frac{\partial\eta}{\partial u} - \frac{\partial^2\tau}{\partial x^2}\right)\Theta_t + \left(\frac{\partial^2\eta}{\partial u^2} - 2\frac{\partial^2\xi}{\partial x\partial u}\right)\Theta_x^2$$

$$+ \left(\frac{\partial\xi}{\partial u} - 2\frac{\partial^2\tau}{\partial x\partial u}\right)\Theta_x\Theta_t + \left(\frac{\partial\tau}{\partial u}\right)\Theta_t^2 - \frac{\partial^2\xi}{\partial u^2}\Theta_x^3$$

$$- \frac{\partial^2\tau}{\partial u^2}\Theta_x^2\Theta_t + \left(\frac{\partial\eta}{\partial u} - 2\frac{\partial\xi}{\partial x}\right)\Theta_{xx} - 2\frac{\partial\tau}{\partial x}\Theta_{xt}$$

$$- 3\frac{\partial\xi}{\partial u}\Theta_{xx}\Theta_x - \frac{\partial\tau}{\partial u}\Theta_{xx}\Theta_t - 2\frac{\partial\tau}{\partial u}\Theta_{xt}\Theta_x \quad . \tag{2.7-3}$$

Since $u = \Theta(x,t)$ satisfies (2.7-1), substituting Θ_t for Θ_{xx} in (2.7-3) the expression simplifies to:

$$\eta_{xx} - \eta_t = \left(\frac{\partial^2\eta}{\partial x^2} - \frac{\partial\eta}{\partial t}\right) + \left(2\frac{\partial^2\eta}{\partial x\partial u} - \frac{\partial^2\xi}{\partial x^2} + \frac{\partial\xi}{\partial t}\right)\Theta_x$$

$$+ \left(\frac{\partial\tau}{\partial t} - 2\frac{\partial\xi}{\partial x} - \frac{\partial^2\tau}{\partial x^2}\right)\Theta_t + \left(\frac{\partial^2\eta}{\partial u^2} - 2\frac{\partial^2\xi}{\partial x\partial u}\right)\Theta_x^2$$

$$+ \left(-2\frac{\partial\xi}{\partial u} - 2\frac{\partial^2\tau}{\partial x\partial u}\right)\Theta_x\Theta_t - \frac{\partial^2\xi}{\partial u^2}\Theta_x^3 - \frac{\partial^2\tau}{\partial u^2}\Theta_x^2\Theta_t$$

$$- 2\frac{\partial\tau}{\partial x}\Theta_{xt} - 2\frac{\partial\tau}{\partial u}\Theta_{xt}\Theta_x \quad . \tag{2.7-4}$$

(2.7-1) is <u>invariant</u> under (2.7-2) iff $\eta_{xx} - \eta_t = 0$ for all x, t and any solution $u = \Theta(x,t)$.

The <u>classical group,</u> henceforth called the group, of the heat equation corresponds to equating to zero the coefficients of terms with the same derivatives of Θ, i.e., the coefficients of $\Theta_x\Theta_{tx}$, $\Theta_t\Theta_x$, Θ_x , Θ_{tx} , Θ_t^2 , Θ_t , Θ_x and the remaining terms not involving derivatives of Θ in (2.7-4), then solving the resulting partial differential equations for ξ, η, τ. This is clearly a sufficient but not a necessary condition for finding similarity solutions (cf. Bluman, Bluman and Cole). However for any system of partial differential equations, linear or non-linear, as mentioned in §2.1, the group arises from solving a set of linear equations for the infinitesimals. Successively equating to zero the coefficients of $\Theta_x\Theta_{tx}$, $\Theta_t\Theta_x$ and Θ_x^2 in (2.7-4), we find that

$$\left.\begin{aligned} \frac{\partial\tau}{\partial u} &= 0 \\ \frac{\partial\xi}{\partial u} &= 0 \\ \frac{\partial^2\eta}{\partial u^2} &= 0 \end{aligned}\right\} \qquad (2.7\text{-}5)$$

$$\left.\begin{aligned} \eta &= f(x,t)u + g(x,t) \\ \xi &= X(x,t) \\ \tau &= T(x,t) \end{aligned}\right\} \qquad (2.7\text{-}6)$$

where f, g, X and T are arbitrary functions of x and t.

Then successively equating to zero the coefficients of Θ_{tx}, Θ_t, Θ_x and the remaining terms, we are led to:

$$T = T(t) \qquad (2.7\text{-}7)$$

$$2\frac{\partial X}{\partial x} - T'(t) = 0 \qquad (2.7\text{-}8)$$

$$\frac{\partial X}{\partial t} - \frac{\partial^2 X}{\partial x^2} + 2\frac{\partial f}{\partial x} = 0 \qquad (2.7\text{-}9)$$

$$\frac{\partial^2 f}{\partial x^2} - \frac{\partial f}{\partial t} = 0 \qquad (2.7\text{-}10)$$

$$\frac{\partial^2 g}{\partial x^2} - \frac{\partial g}{\partial t} = 0 \quad . \qquad (2.7\text{-}11)$$

We note that $g(x,t)$ is any solution to (2.7-1). At first we shall only consider the subgroup for which $g(x,t) = 0$.

Problem 2.7-1:

Solve (2.7-8,9,10) and show that the group of the heat equation is:

$$\begin{aligned} \xi &= \kappa + \delta t + \beta x + \gamma x t \\ \tau &= \alpha + 2\beta t + \gamma t^2 \\ f &= -\gamma\left[\frac{x^2}{4} + \frac{t}{2}\right] - \frac{\delta x}{2} + \lambda \end{aligned} \qquad (2.7\text{-}12)$$

where α, β, γ, δ, κ, λ are six arbitrary parameters which can be complex numbers.

In (x,t) space, the group (2.7-12) is a subgroup of the projective group (cf §1.7). All of the parameters, except for γ, individually represent "trivial" transformations of (x,t) space. κ represents translation in x, α translation in t, δ signifies Galilean invariance, and β denotes stretching invariance (cf §2.4). (2.7-12), in effect, is a six-parameter Lie group of transformations leaving invariant (2.7-1).

In (x,t)-space γ represents invariance under the one-parameter group of projective transformations (cf §1.3)

$$\begin{aligned} x^* &= \frac{x}{1 - \varepsilon t} \\ t^* &= \frac{t}{1 - \varepsilon t} \end{aligned} \qquad (2.7\text{-}13)$$

The Lie algebra constructed from the infinitesimal operators

Let $\mathfrak{X}_i$, $i = 1, 2, \ldots, 6$ represent the infinitesimal operators[4] corresponding to the parameters κ, α, λ, β, γ, and δ respectively.

---

(4) Cf. §1.5 and p. 203 (2.6-10) of §2.6.

Then

$$\mathfrak{X}_1 = \frac{\partial}{\partial x}$$

$$\mathfrak{X}_2 = \frac{\partial}{\partial t}$$

$$\mathfrak{X}_3 = u \frac{\partial}{\partial u}$$

$$\mathfrak{X}_4 = x \frac{\partial}{\partial x} + 2t \frac{\partial}{\partial t} \tag{2.7-14}$$

$$\mathfrak{X}_5 = xt \frac{\partial}{\partial x} + t^2 \frac{\partial}{\partial t} - [\frac{x^2}{4} + \frac{t}{2}]u \frac{\partial}{\partial u}$$

$$\mathfrak{X}_6 = t \frac{\partial}{\partial x} - \frac{x}{2} u \frac{\partial}{\partial u}$$

For any m-parameter Lie group the infinitesimal operators form an m-dimensional Lie algebra, $\mathscr{M}$. A Lie algebra, $\mathscr{L}$, is a vector space over some field $\mathscr{F}$ with in addition a law of combination (called commutation) $\otimes$ connecting elements of $\mathscr{L}$ such that the following axioms are satisfied:

(i) If $\mathfrak{X} \varepsilon \mathscr{L}$, $\mathfrak{Y} \varepsilon \mathscr{L}$, then $\mathfrak{X} \otimes \mathfrak{Y} \varepsilon \mathscr{L}$. (closure property).

(ii) $\mathfrak{X} \otimes \mathfrak{Y} = - \mathfrak{Y} \otimes \mathfrak{X}$

(iii) If $\mathfrak{X}$, $\mathfrak{Y}$, $\mathfrak{Z}$ $\varepsilon \mathscr{L}$, then

$$\mathfrak{X} \otimes (\mathfrak{Y} \otimes \mathfrak{Z}) + \mathfrak{Y} \otimes (\mathfrak{Z} \otimes \mathfrak{X}) + \mathfrak{Z} \otimes (\mathfrak{X} \otimes \mathfrak{Y}) = 0$$

(Jacobi's identity).

An elementary example of a Lie algebra is the set of vectors in R^3 (three dimensional Euclidean space) with $\otimes$ the vector product. The infinitesimal operators of a Lie group of transformations form a vector space over the field of real or complex numbers. In order for this vector space to become a Lie algebra, $\mathscr{M}$, the commutation operation, [,] , is introduced.

If $\mathfrak{X}, \mathfrak{Y} \in \mathcal{M}$ then

$$\mathfrak{X} \otimes \mathfrak{Y} = [\mathfrak{X}, \mathfrak{Y}] = \mathfrak{X}\mathfrak{Y} - \mathfrak{Y}\mathfrak{X}$$

where $[\mathfrak{X}, \mathfrak{Y}]$ is the commutator of the infinitesimal operator $\mathfrak{X}$ with the infinitesimal operator $\mathfrak{Y}$. It is left as an exercise for the reader to verify that the axioms (i) to (iii) are satisfied.

For every Lie algebra a commutator table can be established. For the infinitesimal operators (2.7-14) the following is the commutator table[(5)]:

	$\mathfrak{X}_1$	$\mathfrak{X}_2$	$\mathfrak{X}_3$	$\mathfrak{X}_4$	$\mathfrak{X}_5$	$\mathfrak{X}_6$
$\mathfrak{X}_1$	0	0	0	$\mathfrak{X}_1$	$\mathfrak{X}_6$	$-\frac{1}{2}\mathfrak{X}_3$
$\mathfrak{X}_2$	0	0	0	$2\mathfrak{X}_2$	$\mathfrak{X}_4-\frac{1}{2}\mathfrak{X}_3$	$\mathfrak{X}_1$
$\mathfrak{X}_3$	0	0	0	0	0	0
$\mathfrak{X}_4$	$-\mathfrak{X}_1$	$-2\mathfrak{X}_2$	0	0	$2\mathfrak{X}_5$	$\mathfrak{X}_6$
$\mathfrak{X}_5$	$-\mathfrak{X}_6$	$-\mathfrak{X}_4+\frac{1}{2}\mathfrak{X}_3$	0	$-2\mathfrak{X}_5$	0	0
$\mathfrak{X}_6$	$\frac{1}{2}\mathfrak{X}_3$	$-\mathfrak{X}_1$	0	$-\mathfrak{X}_6$	0	0

Hence we see that the Lie algebra of (2.7-14) is generated by the operators $\mathfrak{X}_1$, $\mathfrak{X}_2$, and $\mathfrak{X}_5$. This follows from

$[\mathfrak{X}_1, \mathfrak{X}_5] = \mathfrak{X}_6$, generating $\mathfrak{X}_6$

$[\mathfrak{X}_1, \mathfrak{X}_6] = -\frac{1}{2}\mathfrak{X}_3$, generating $\mathfrak{X}_3$

$[\mathfrak{X}_2, \mathfrak{X}_5] = \mathfrak{X}_4 - \frac{1}{2}\mathfrak{X}_3$, generating $\mathfrak{X}_4$

(5) The result of $[\mathfrak{X}_i, \mathfrak{X}_j]$ corresponds to the intersection of the ith row and the j-th column.

i.e. from invariance with respect to parameters κ, α and γ we know that the heat equation must be invariant with respect to the one-parameter groups corresponding to β, δ and λ. Note that it is impossible to generate the operators $\mathfrak{X}_2$ and $\mathfrak{X}_5$ from knowledge of invariance with respect to the operators $\mathfrak{X}_1$, $\mathfrak{X}_3$, $\mathfrak{X}_4$ and $\mathfrak{X}_6$.

Construction of the "general" similarity solution of the heat equation

The characteristic equations used to find the general similarity[6] solution of the heat equation (cf. §2.1) are:

$$\frac{dx}{\xi} = \frac{dt}{\tau} = \frac{du}{fu} \tag{2.7-15}$$

where ξ, τ and f are given by (2.7-12).

The most general "moving" boundary (represented by the similarity variable ζ) invariant under the group (2.7-12) is found by solving the first order differential equation corresponding to the first equality in (2.7-15). Four cases are distinguished.

Case I $\quad \beta^2 \neq \alpha\gamma \leftrightarrow \zeta = \dfrac{x - (At + B)}{\sqrt{\alpha + 2\beta t + \gamma t^2}}$

where

$$A = \frac{\kappa\gamma - \delta\beta}{\alpha\gamma - \beta^2}, \quad B = \frac{\kappa\beta - \delta\alpha}{\alpha\gamma - \beta^2}.$$

Case II $\quad \beta^2 = \alpha\gamma,\ \gamma \neq 0 \leftrightarrow \zeta = \left[x + \delta + \dfrac{\kappa - \delta\beta}{2(t + \beta)}\right] \cdot \dfrac{1}{t + \beta}$.

Case III $\quad \beta^2 = \alpha\gamma,\ \beta = \gamma = 0,\ \alpha \neq 0 \leftrightarrow \zeta = x - \dfrac{\delta t^2}{2} - \kappa t$.

Case IV $\quad \alpha = \beta = \gamma = 0,\ \delta \neq 0 \leftrightarrow \zeta = t$.

In all of these cases the group leaves invariant boundaries of

---

(6) A similarity solution is a solution obtained from group invariance. The general similarity solution is the most general similarity solution obtained directly from group invariance.

the form ζ = constant. For example - say we have a problem requiring us to leave invariant a fixed boundary $x = x_o$ and a moving boundary $x = x_1 + ct$. This would correspond to Case II with $\kappa = \delta\beta$, $\gamma = 1$,

$$\Rightarrow \quad \begin{cases} \delta = -x_o \\ \beta = \dfrac{x_1 - x_o}{c} \\ \alpha = \left(\dfrac{x_1 - x_o}{c}\right)^2 \\ \kappa = -\dfrac{x_o}{c}(x_1 - x_o) \end{cases} \quad . \tag{2.7-16}$$

In this case $\zeta = \dfrac{x + \delta}{t + \beta}$

where $\zeta = 0 \leftrightarrow x = x_o$

$\zeta = c \leftrightarrow x = x_1 + ct$.

In general it is easier to work directly with the group (2.7-12) in order to determine whether a particular boundary curve is left invariant. In the above example invariance of $x = x_o$ and $x = x_1 + ct$, respectively, implies that

$$\xi(x = x_o, t) = 0$$

and

$$\xi(x = x_1 + ct, t) = c\tau(x = x_1 + ct, t) \quad . \tag{2.7-17}$$

Hence

$$\kappa + \delta t + \beta x_o + \gamma x_o t = 0$$

and

$$\kappa + \delta t + \beta(x_1 + ct) + \gamma t(x_1 + ct) = c(\alpha + 2\beta t + \gamma t^2)$$

$\Rightarrow$ the result of (2.7-16) for the respective parameters (Case II).

Corresponding to the four distinguished cases we now construct the similarity solutions obtained by solving the characteristic equations (2.7-14):

Case I

The similarity form for the solution is

$$u = \Theta(x,t) = F(\zeta)\cdot\frac{1}{(\alpha+2\beta t+\gamma t^2)^{1/4}}\cdot\left(\frac{\gamma t+\beta-C}{\gamma t+\beta+C}\right)^{\rho}\cdot\exp\{-\frac{t}{4}(A^2+\gamma\zeta^2)+\frac{A\zeta}{2}\sqrt{\alpha+2\beta t+\gamma t^2}\}$$

where

$$C = \sqrt{\beta^2 - \alpha\gamma}$$

and

$$\rho = \frac{1}{2C}\{\frac{\beta}{2} + \lambda + \frac{1}{4\gamma}(\delta^2 - A^2C^2)\} \quad . \tag{2.7-18}$$

Substituting (2.7-18) into the heat equation we obtain the following ordinary differential equation for $F(\zeta)$:

$$F'' + \beta\zeta F' + (D\zeta^2 + E)F = 0 \tag{2.7-19}$$

where a prime denotes differentiation with respect to ζ and

$D = \frac{\alpha\gamma}{4}$,

$$E = \frac{A^2C^2}{4\gamma} - (\lambda + \frac{\delta^2}{4})$$

$$= - 2C\rho + \frac{\beta}{2} \quad .$$

Now let $z = \zeta\sqrt{C}$ and $F(\zeta) = G(z)e^{-\frac{\beta\zeta^2}{4}}$. Then (2.7-19) becomes

$$\frac{d^2G}{dz^2} + [\frac{1}{2} + \nu - \frac{1}{4}z^2]G = 0 \tag{2.7-19'}$$

where

$$\nu = \frac{E - \frac{\beta}{2}}{C} - \frac{1}{2} = - 2\rho - \frac{1}{2} \quad .$$

This equation is of a standard confluent hypergeometric type whose solution can be expressed in terms of parabolic cylinder functions. Any two of

$$D_\nu(z),\ D_\nu(-z),\ D_{-\nu-1}(iz),\ D_{-\nu-1}(-iz)$$

are linearly independent solutions of (2.7-19'). Their properties are well known.

$$D_\nu(z) = 2^{\frac{\nu}{2}} e^{-\frac{z^2}{4}} \cdot [\frac{\Gamma(\frac{1}{2})}{\Gamma(\frac{1}{2}-\frac{1}{2}\nu)} {}_1F_1(-\tfrac{1}{2}\nu;\tfrac{1}{2};\tfrac{1}{2}z^2)$$
$$+ 2^{-\frac{1}{2}} z \frac{\Gamma(-\frac{1}{2})}{\Gamma(-\frac{1}{2}\nu)} {}_1F_1(\tfrac{1}{2}-\tfrac{1}{2}\nu;\tfrac{3}{2};\tfrac{1}{2}z^2)] \quad .$$

For integer values $\nu = n = 0,1,2,\ldots,$ the solutions are expressed by (orthogonal) Hermite polynomials

$$D_n(z) = e^{-\frac{z^2}{4}} He_n(z) \ ;$$
$$He_n = (-1)^n e^{\frac{z^2}{2}} \frac{d^n}{dz^n} e^{-\frac{z^2}{2}}$$

$$He_o = 1, \ He_1 = z, \ He_2 = z^2 - 1, \text{ etc.}$$

While

$$D_{-1}(z) = e^{\frac{z^2}{4}} (2\pi)^{-\frac{1}{2}} \operatorname{erfc}\,(2^{-\frac{1}{2}} z), \ \ldots$$

The series representation of $D_\nu(z)$ shows the behavior for small z while asymptotically for $|z| >> |\nu|, 1$

$$D_\nu(z) = \begin{cases} z^\nu e^{-\frac{z^2}{4}} [1 + O(z^{-2})], \ |\arg z| < \frac{3\pi}{4} \quad . \\ z^\nu e^{-\frac{z^2}{4}} [1 + O(z^{-2})] - \frac{(2\pi)^{1/2}}{\Gamma(-\nu)} e^{\nu\pi i} z^{-\nu-1} e^{\frac{z^2}{4}} \\ \cdot [1 + O(z^{-2})], \ \frac{5\pi}{4} > \arg z > \frac{\pi}{4} \quad . \\ z^\nu e^{-\frac{z^2}{4}} [1 + O(z^{-2})] - \frac{(2\pi)^{1/2}}{\Gamma(-\nu)} e^{-2\pi i} z^{-\nu-1} e^{\frac{z^2}{4}} \\ \cdot [1 \pm O(z^{-2})], \ -\frac{\pi}{4} > \arg z > -\frac{5\pi}{4} \quad . \end{cases}$$

The parabolic cylinder functions are <u>entire</u> functions of z .

Case II

The similarity form for u is

$$\Theta = \frac{F(\zeta)}{\sqrt{t + \beta}} \exp \left[\frac{L^2}{12(t + \beta)^3} + \frac{M}{t + \beta} - \frac{L\zeta}{2(t + \beta)} - \frac{\zeta^2(t + \beta)}{4}\right] \qquad (2.7\text{-}20)$$

where $L = \frac{\kappa - \delta\beta}{2}$

$$M = -\left(\frac{\beta}{2} + \lambda + \frac{\delta^2}{4}\right)$$

and $F(\zeta)$ satisfies

$$\frac{d^2F}{d\zeta^2} - [L\zeta - M]F = 0 \quad . \qquad (2.7\text{-}21)$$

Letting $z = L^{1/3}\zeta$ and $F(\zeta) = G(z)$, (2.7-21) becomes

$$\frac{d^2G}{dz^2} - (z - \nu)G = 0 \qquad (2.7\text{-}21')$$

where

$$\nu = \frac{M}{L^{2/3}} \quad .$$

The Airy functions $Ai(z - \nu)$ and $Bi(z - \nu)$ are linearly independent solutions of (2.7-21'). These can be represented in term of Bessel functions of order $\frac{1}{3}$.

If $$w = \frac{2}{3}z^{3/2} ,$$

$$Ai(z) = \pi^{-1}\sqrt{\frac{z}{3}}K_{\frac{1}{3}}(w) ,$$

$$Ai(-z) = \frac{1}{3}\sqrt{z}[J_{\frac{1}{3}}(w) + J_{-\frac{1}{3}}(w)] ,$$

$$Bi(z) = \sqrt{\frac{z}{3}}[I_{-\frac{1}{3}}(w) + I_{\frac{1}{3}}(w)] ,$$

$$Bi(-z) = \sqrt{\frac{z}{3}}[J_{-\frac{1}{3}}(w) - J_{\frac{1}{3}}(w)] .$$

Case III

Here the similarity form is

$$u = \Theta(x,t) = F(\zeta)\, \exp[-\frac{\delta^2}{12}t^3 - \frac{\delta\kappa t^2}{4} + \lambda t - \frac{\delta}{2}\zeta t] \tag{2.7-22}$$

with $F(\zeta)$ satisfying

$$\frac{d^2F}{d\zeta^2} + \kappa\frac{dF}{d\zeta} + [\frac{\delta}{2}\zeta - \lambda]F = 0 \quad . \tag{2.7-23}$$

Letting $z = -(\frac{\delta}{2})^{1/3}\zeta$ and $F(\zeta) = G(z)e^{-\frac{\kappa}{2}\zeta}$, $\delta \neq 0$ (7)

(2.7-23) becomes

$$\frac{d^2G}{dz^2} - (z - \nu)G = 0 \tag{2.7-23'}$$

where

$$\nu = -(\frac{\kappa^2}{4} + \lambda)(\frac{2}{\delta})^{2/3}$$

which is the same equation as (2.7-21') .

Case IV

Here the resulting similarity form for the solution is

$$u = \Theta(x,t) = F(t)e^{\frac{x}{4(t+\kappa)}[\lambda - x]} \tag{2.7-24}$$

with $F(t)$ satisfying the first order equation

$$\frac{dF}{dt} = \frac{\lambda^2 - 8(t+\kappa)}{16(t+\kappa)^2}\, F(t) \tag{2.7-25}$$

$$\Rightarrow \quad \Theta(x,t) = \frac{P}{\sqrt{t+\kappa}}\, e^{-\frac{(x - \frac{\lambda}{2})^2}{4(t+\kappa)}} \tag{2.7-26}$$

---

(7) If in addition $\kappa = 0$, the classical separation of variables solution is included here, $u = e^{\lambda t}\{A\cos\sqrt{-\lambda}x + B\sin\sqrt{-\lambda}x\}$

with P an arbitrary constant. This is the familiar source solution.

It should be noted that the parameters of the Lie group (2.7-12) can be complex. Since the heat equation is linear the real and imaginary parts of the resulting complex similarity solutions are also solutions.

Examples where $g(x,t) \neq 0$

In general for a Lie group of the form (2.7-6), the characteristic equations to obtain the similarity form $u = \Theta(x,t)$ of the solution are:

$$\frac{dx}{\xi(x,t)} = \frac{dt}{\tau(x,t)} = \frac{d\Theta}{f(x,t)\Theta + g(x,t)} \tag{2.7-27}$$

Integrating the first equality of (2.7-27) we obtain the similarity variable

$$\zeta(x,t) = \text{const.} \quad (\Rightarrow x = h(\zeta,t)) \tag{2.7-28}$$

Θ then satisfies the first order linear differential equation (ζ = const.)

$$\frac{d\Theta}{dt} = \frac{f(h(\zeta,t),t)\Theta}{\tau(h(\zeta,t),t)} + \frac{g(h(\zeta,t),t)}{\tau(h(\zeta,t),t)} \quad . \tag{2.7-29}$$

To solve this equation we would first let $\Theta_P(t,\zeta)$ be a particular nontrivial solution of the homogeneous part of (2.7-29). Such a solution is easily generated by setting $F(\zeta) = \text{const.} = 1$, say, in the similarity form generated when $g(x,t) = 0$. Then the similarity form for the invariant solution corresponding to (2.7-27) is :

$$\Theta(x,t) = \Theta_P(t,\zeta)H(t) \tag{2.7-30}$$

where (a arbitrary)

$$H(t) = \int_a^t \frac{g(h(\zeta,t^*),t^*)dt^*}{\tau(h(\zeta,t^*),t^*)\Theta_p(t^*,\zeta)} + F(\zeta) \tag{2.7-31}$$

and $F(\zeta)$ is an arbitrary function of ζ.

Substituting (2.7-30) into the heat equation an ordinary differential equation is obtained for $F(\zeta)$.

Problem 2.7-1.

Show that if $u = g(x,t)$ is a solution to the heat equation, then so are

(i) $(\xi = 0,\ \tau = 1,\ f = 0)$

$$u = \int_a^t g(x,t^*)dt^* + F(x)$$

where

$$F''(x) = g(x,a) \quad ,$$

(ii) $(\xi = 1,\ \tau = 0,\ f = 0)$

$$u = \int_a^x g(x^*,t)dx^* + F(t)$$

where

$$F'(t) = g_x(a,t) \quad .$$

(iii) $(\xi = \tau = 1,\ f = 0)$

$$u = \int_a^x g(x^*,x^* + \zeta)dx^* + F(\zeta)$$

where

$$\zeta = t - x$$

and

$$F''(\zeta) - F'(\zeta) = -[g_x(a,a + \zeta) - g_t(a,a + \zeta)]$$

Problem 2.7-2

Find the similarity solution of the heat equation corresponding to the case where $g(x,t)$ is any solution and $f = 1$, $\xi = x$, $\tau = 2t$.

Problem 2.7-3

Consider the heat equation in two and three dimensions.

(i) Find the nine-parameter group leaving invariant

$$u_{xx} + u_{yy} - u_t = 0 .$$

(ii) Find the thirteen-parameter group leaving invariant

$$u_{xx} + u_{yy} + u_{zz} - u_t = 0 .$$

Summary.

Using infinitesimal transformations we have shown that the heat equation is invariant under a six-parameter group of transformations. Two of these parameters correspond to nontrivial symmetries since these symmetries could not have been guessed through "inspectional" analysis. The general similarity solution corresponding to this six-parameter group has been obtained. In the next section we will consider the application of group methods to boundary value problems for the one-dimensional heat equation.

2.8. Applications of the General Similarity Solution of the Heat Equation

In this section we will consider some particular applications of the similarity solutions of §2.7 to boundary value problems.[1]

Fundamental solutions of the heat equation

(i) infinite bar; invariance under a multi-parameter group.

---

(1) G. W. Bluman, Applications of the General Similarity Solution of the Heat Equation to Boundary Value Problems, Quart. of A.Ma., 1974.

Because of invariance under translations in x, for an infinite bar we can assume that a source is located at $x = 0$, represented by

$$u(x,0) = \delta(x) \tag{2.8-1}$$

where $\delta(x)$ is the Dirac delta function. In this case $t = 0$ is the only boundary to be left invariant. We recall that the group of the heat equation is:

$$\begin{aligned} \xi(x,t) &= \kappa + \delta t + \beta x + \gamma xt \\ \tau(x,t) &= \tau(t) = \alpha + 2\beta t + \gamma t^2 \\ f(x,t) &= -\gamma\left[\frac{x^2}{4} + \frac{t}{2}\right] - \frac{\delta x}{2} + \lambda \quad . \end{aligned} \tag{2.8-2}$$

We now find the subgroup of (2.8-2) leaving invariant the boundary condition (2.8-1) prescribed on the curve $t = 0$. Invariance of $t = 0$ implies that

$$t^* = t + \varepsilon\tau(t) = 0 \qquad \text{when} \quad t = 0 \ ,$$

i.e., $\tau(0) = 0$. Hence $\alpha = 0$.

Invariance of (2.8-1) implies that

$$u^*(x,0) = \delta(x^*) \quad . \tag{2.8-3}$$

Expanding both sides of (2.8-3) about $\varepsilon = 0$, we find that

$$\begin{aligned} u^*(x,0) &= u(x,0) + \varepsilon f(x,0)u(x,0) + O(\varepsilon^2) \ , \\ \delta(x^*) &= \delta(x) + \varepsilon\xi(x,0)\delta'(x) + O(\varepsilon^2) \quad . \end{aligned} \tag{2.8-4}$$

Thus invariance of the source condition requires that

$$f(x,0)\delta(x) = \xi(x,0)\delta'(x) \quad . \tag{2.8-5}$$

Formally $\qquad x\delta'(x) = -\delta(x)$

$$=> \qquad A(x)\delta'(x) = -\frac{A(x)}{x}\delta(x)$$

$$= -A'(x)\delta(x) \qquad \text{if} \quad A(0) = 0 \quad .$$

Moreover $\quad B(x)\delta(x) = 0 \quad$ if $\quad B(0) = 0$.

Hence the invariance boundary condition (2.8-5) is satisfied if

$$\begin{aligned} \xi(0,0) &= 0 \\ f(0,0) &= -\xi_x(0,0) \quad . \end{aligned} \qquad (2.8\text{-}6)$$

Thus $\kappa = 0$ and $\lambda = -\beta$. So that finally a three-parameter sub-group

$$\begin{aligned} \xi &= \delta t + \beta x + \gamma x t \\ \tau &= 2\beta t + \gamma t^2 \\ f &= -\gamma[\frac{x^2}{4} + \frac{t}{2}] - \frac{\delta x}{2} - \beta \end{aligned} \qquad (2.8\text{-}7)$$

leaves invariant the heat equation and (2.8-1). Each of these parameters can be taken in turn to generate a similarity form for the solution. Equating the functional forms corresponding to any two of these parameters we solve the resulting functional equation to obtain a solution containing some arbitrary constant which is computed by imposing the initial source condition. Thus invariance under a two-parameter group means that no further use needs to be made of the given partial differential equation (provided it contains precisely two independent variables) as in the case of invariance under a one-parameter group[2].

<u>Case I</u>. $\beta = \gamma = 0$, $\delta = 1$

The characteristic equations are

---

[2] G. W. Bluman, Similarity solutions of the one-dimensional Fokker-Planck equation, Int. J. Non-Lin. Mech., 6, pp. 143-153, 1971.

$$\frac{dx}{t} = \frac{dt}{0} = \frac{du}{-\frac{x}{2}u} \tag{2.8-8}$$

=> $\zeta_1 = t$ is the similarity variable and the similarity form for the solution is

$$u = \Theta(x,t) = F_1(\zeta_1)e^{-\frac{x^2}{4\zeta_1}} \tag{2.8-9}$$

where F_1 is an arbitrary function of ζ_1.

Case II. $\delta = \gamma = 0$, $\beta = 1$ (the "usual" group used to obtain the source solution)

Here the characteristic equations are

$$\frac{dx}{x} = \frac{dt}{2t} = \frac{du}{-u} \quad . \tag{2.8-10}$$

The similarity variable is $\zeta_2 = \frac{x}{\sqrt{t}}$ with similarity form

$$u = \Theta(x,t) = \frac{1}{\sqrt{t}} F_2(\zeta_2) \tag{2.8-11}$$

where F_2 is an arbitrary function of ζ_2.

Equating the functional similarity forms we obtain a functional equation to be solved for either F_1 or F_2 :

$$F_1(\zeta_1)e^{-\frac{x^2}{4\zeta_1}} = \frac{1}{\sqrt{t}} F_2(\zeta_2) \tag{2.8-12}$$

$$\Rightarrow \quad \sqrt{\zeta_1}\, F_1(\zeta_1) = e^{\frac{\zeta_2^2}{4}} F_2(\zeta_2) \tag{2.8-13}$$

since $$t = \zeta_1$$

and $$\frac{x^2}{\zeta_1} = \zeta_2^2 \quad .$$

Now using ζ_1 and ζ_2 as our new variables we easily see that (2.8-13) is satisfied iff $\sqrt{\zeta_1}\, F_1(\zeta_1) = \text{constant} = c$, say.

$$\therefore \quad F_1(\zeta_1) = \frac{c}{\sqrt{t}} \tag{2.8-14}$$

$$\Rightarrow \quad \Theta(x,t) = \frac{c}{\sqrt{t}}\, e^{-\frac{x^2}{4t}} \quad . \tag{2.8-15}$$

The source condition => $c = \dfrac{1}{2\sqrt{\pi}}$.

It can be shown (without going through calculations) that the one-parameter (γ) subgroup leads to the same solution.

Say

$$\mathfrak{X} = \xi \frac{\partial}{\partial x} + \tau \frac{\partial}{\partial t} + \eta \frac{\partial}{\partial u}$$

is the infinitesimal operator of some one-parameter group and $u - \Theta(x,t) = 0$ is a similarity solution corresponding to invariance under $\mathfrak{X}$. Then (cf. §2.1)

$$\mathfrak{X}\{u - \Theta(x,t)\} = 0 \tag{2.8-16}$$

(2.8-16) simply states that $\Theta(x,t)$ satisfies the invariant surface condition (2.1-8) corresponding to invariance under $\mathfrak{X}$.

Now consider the infinitesimal operators corresponding respectively to the parameters δ, β and γ of the group (2.8-7) leaving invariant the heat equation and the boundary condition (2.8-1), namely

$$\begin{aligned} \mathfrak{X}_1 &= t \frac{\partial}{\partial x} - \frac{1}{2} xu \frac{\partial}{\partial u} \\ \mathfrak{X}_2 &= x \frac{\partial}{\partial x} + 2t \frac{\partial}{\partial t} - u \frac{\partial}{\partial u} \\ \mathfrak{X}_3 &= xt \frac{\partial}{\partial x} + t^2 \frac{\partial}{\partial t} - \frac{1}{4}(x^2 + 2t)u \frac{\partial}{\partial u} \quad . \end{aligned} \tag{2.8-17}$$

We note that

$$\mathfrak{X}_3 = \frac{t}{2}\mathfrak{X}_2 + \frac{x}{2}\mathfrak{X}_1 \quad . \tag{2.8-18}$$

Let

$$\Theta_o(x,t) = \frac{1}{\sqrt{4\pi t}} e^{-\frac{x^2}{4t}} \quad . \tag{2.8-19}$$

Then $u - \Theta_o(x,t) = 0$ is the similarity solution constructed from invariance under the two parameter group with operators $\mathfrak{X}_1$ and $\mathfrak{X}_2$, i.e., $u = \Theta_o(x,t)$ satisfies the heat equation, the boundary condition (2.8-1), and moreover

$$\mathfrak{X}_1\{u - \Theta_o(x,t)\} = 0$$

and

$$\mathfrak{X}_2\{u - \Theta_o(x,t)\} = 0 \quad .$$

Thus for any functions $\lambda_1(x,t,u)$, $\lambda_2(x,t,u)$:

$$[\lambda_1(x,t,u)\ \mathfrak{X}_1 + \lambda_2(x,t,u)\ \mathfrak{X}_2]\{u - \Theta_o(x,t)\} = 0$$

if $u = \Theta_o(x,t)$. In particular $\mathfrak{X}_3\{u - \Theta_o(x,t)\} = 0$. Thus (2.8-15) is <u>a</u> similarity form corresponding to invariance under the one-parameter (γ) group.(3)

<u>Problem 2.8-1</u>

The source solution (2.8-15) was obtained using the two-parameter (δ,β) subgroup. Show that (δ,γ) and (β,γ) form two-parameter subgroups and obtain the source solution from invariance under each of these subgroups.

---

(3) Thus we see that if a p.d.e. and associated boundary conditions in ℓ variables having a unique solution is invariant under an m-parameter Lie group G_m then from this invariance the number of variables is reduced by $k \leq \min\{\ell - 1, m\}$ where k is the dimension of the Lie algebra $\mathscr{M}$ corresponding to G_m over the field of complex valued functions of the ℓ variables. Recall that m is the dimension of $\mathscr{M}$ over the field of complex numbers.

(ii) semi-infinite bar

In this case the heat is distributed over the region $x > 0$ with the end $x = 0$ either insulated $(u_x = 0)$ or kept at a fixed temperature $u = 0$, say. The governing equations are:

$$u_{xx} = u_t \tag{2.8-20}$$

with boundary conditions

$$u(x,0) = \delta(x - x_o) \quad , \quad x_o > 0 \tag{2.8-21}$$

$$Au_x(0,t) + Bu(0,t) = 0 \tag{2.8-22}$$

where either $A = 0$ or $B = 0$.

This problem can be easily solved by using images of the source solution for an infinite bar. Let

$$G(x,t) = \frac{1}{\sqrt{4\pi t}} e^{-\frac{x^2}{4t}} \tag{2.8-23}$$

represent the source solution for an infinite bar with the source located at $x = 0$. If $A = 0$ we solve (2.8-20,21,22) by placing a source at $x = x_o$ and a source (negative source) at $x = -x_o$, i.e., the solution is

$$u = G(x - x_o,t) - G(x + x_o,t) \quad . \tag{2.8-24}$$

If $B = 0$ we place sources at $x = \pm x_o$

$$\Rightarrow \quad u = G(x - x_o,t) + G(x + x_o,t) \quad . \tag{2.8-25}$$

Instead of using images we will now proceed directly using the group invariance technique.

As before invariance of $t = 0$ and (2.8-21) implies that

$$\begin{aligned} \tau(t = 0) &= 0 \\ f(x_o,0) &= -\xi_x(x_o,0) \\ \xi(x_o,0) &= 0 \quad . \end{aligned} \tag{2.8-26}$$

Invariance of (2.8-22) means that $x = 0$ must also be invariant

$$\Rightarrow \quad \xi(0,t) = 0 \quad . \tag{2.8-27}$$

In addition if $B = 0$ then

$$u_x(0,t) = 0 \quad \Leftrightarrow \quad u^*_{x^*}(0,t) = 0$$

$$\text{iff} \quad f_x(0,t) = 0 \quad . \tag{2.8-28}$$

It turns out that (2.8-28) is satisfied if (2.8-26,27) are satisfied.

The end result is that $\alpha = \kappa = \delta = \beta = 0$ and $\lambda = \frac{\gamma x_o^2}{4}$.
Thus the one-parameter group

$$\begin{aligned} \xi &= xt \\ \tau &= t^2 \\ f &= -\left(\frac{x^2}{4} + \frac{t}{2}\right) + \frac{x_o^2}{4} \end{aligned} \tag{2.8-29}$$

leaves invariant (2.8-20,21,22).

Solving the characteristic equations

$$\frac{dx}{xt} = \frac{dt}{t^2} = \frac{du}{u\left[-\frac{x^2}{4} - \frac{t}{2} + \frac{x_o^2}{4}\right]} \tag{2.8-30}$$

we find that the similarity variable is $\zeta = \frac{x}{t}$ and u has the similarity form

$$u = \Theta(x,t) = \frac{e^{-\frac{(x^2+x_o^2)}{4t}}}{\sqrt{t}} F(\zeta) \quad . \tag{2.8-31}$$

Substituting (2.8-31) into (2.8-20) we see that F satisfies the ordinary differential equation

$$\frac{d^2F}{d\zeta^2} = \frac{x_o^2}{4} F \tag{2.8-32}$$

whose general solution is

$$F(\zeta) = Ce^{\frac{x_o}{2}\zeta} + De^{-\frac{x_o}{2}\zeta} . \tag{2.8-33}$$

Hence

$$u = \frac{1}{\sqrt{t}}[Ce^{-\frac{(x-x_o)^2}{4t}} + De^{-\frac{(x+x_o)^2}{4t}}] , \tag{2.8-34}$$

$$u_x(0,t) = 0 \quad => \quad D = C .$$

$$u(0,t) = 0 \quad => \quad D = -C .$$

$$u(x,0) = \delta(x - x_o) \quad => \quad C = \frac{1}{\sqrt{4\pi}} .$$

So that

$$u = \frac{1}{\sqrt{4\pi t}}[G(x - x_o,t) \pm G(x + x_o,t)] .$$

(iii) <u>finite bar; superposition of similarity solutions</u>

In this case heat is distributed over the region $0 < x < \ell$ with a source at $x = x_o$, $0 < x_o < \ell$. Each end is either insulated or kept at zero temperature. Thus we wish to solve

$$u_{xx} = u_t \tag{2.8-35}$$

with boundary conditions

$$u(x,0) = \delta(x - x_o) , \quad 0 < x_o < \ell \tag{2.8-36}$$

$$Au_x(0,t) + Bu(0,t) = 0 , \quad A = 0 \text{ or } B = 0 \tag{2.8-37}$$

$$Cu_x(\ell,t) + Du(\ell,t) = 0 , \quad C = 0 \text{ or } D = 0. \tag{2.8-38}$$

This problem was solved by the method of images in §2.4. We now proceed using group methods to generate the well-known Fourier series representation of the solution. Since only the trivial subgroup leaves invariant the boundary conditions (2.8-36,37,38) we seek similarity solutions of the heat equation corresponding to invariance of the heat equation and (2.8-37,38). Then we use a linear superposition of such similarity solutions to satisfy the boundary condition (2.8-36). In effect we are now considering the use of groups to solve a problem which is "almost" invariant.

In order to leave invariant (2.8-37,38) we must leave invariant the boundaries $x = 0$ and $x = \ell$. Hence $\kappa = \delta = \beta = \gamma = 0$ with λ and α still arbitrary. We set $\alpha = 1$, relabel λ by $-\lambda^2$. To generate the initial source we will use superposition of the eigensolutions of (2.8-35,37,38) corresponding to the eigenvalues of λ. The characteristic equations are

$$\frac{dx}{0} = \frac{dt}{1} = \frac{du}{-\lambda^2 u} \quad . \qquad (2.8\text{-}39)$$

The similarity variable is $\zeta = x$ and the corresponding similarity form for u is:

$$u = e^{-\lambda^2 t} F_\lambda(x) \qquad (2.8\text{-}40)$$

for each value of λ.

F_λ satisfies the ordinary differential equation

$$\frac{d^2 F_\lambda}{dx^2} = -\lambda^2 F_\lambda \qquad (2.8\text{-}41)$$

$$\Rightarrow \quad F_\lambda = K_\lambda \sin \lambda x + L_\lambda \cos \lambda x \qquad (2.8\text{-}42)$$

where K_λ and L_λ are arbitrary constants. Assuming zero temperature at each end, ($A = C = 0$ in 2.8-37,38) we find that $L_\lambda = 0$ and that the eigenvalues are

$$\lambda = \lambda_n = \frac{n\pi}{\ell} \,, \quad n = 1, 2, \ldots$$

Thus

$$u = \sum_{n=1}^{\infty} K_n \sin \frac{n\pi}{\ell} x \, e^{-\frac{n^2\pi^2}{\ell^2} t} \quad . \qquad (2.8\text{-}43)$$

$$u(x,0) = \delta(x - x_o) \quad \Rightarrow$$

$$K_n = \frac{2}{\ell} \sin \frac{n\pi}{\ell} x_o \,, \quad n = 1, 2, \ldots \qquad (2.8\text{-}44)$$

Rayleigh flow problem

The use of stretching invariance to find the solution of a particular Rayleigh flow problem was discussed in §2.5. We will now show how to find the source solution to the Rayleigh flow problem using a subgroup of the full group of the heat equation. The equation and boundary conditions are:

$$\frac{\partial u}{\partial t} = \nu \frac{\partial^2 u}{\partial y^2}$$

$$u(y,0) = 0 \ , \ y > 0 \ , \tag{2.8-45}$$

$$u(0,t) = \delta(t) \ ,$$

$$u(y \to \infty, t) = 0 \ .$$

Letting $x = \frac{y}{\sqrt{\nu}}$ (2.8-45) becomes

$$\frac{\partial u}{\partial t} = \frac{\partial^2 u}{\partial x^2} \tag{2.8-46}$$

with boundary conditions:

$$u(x,0) = 0 \ , \quad x > 0 \ , \tag{2.8-47}$$

$$u(0,t) = \delta(t) \ , \tag{2.8-48}$$

$$u(x \to \infty, t) = 0 \ . \tag{2.8-49}$$

Invariance of $t = 0$ and $x = 0 \Rightarrow \quad \tau(t = 0) = \xi(0,t) = 0$

$$\Rightarrow \qquad \alpha = \kappa = \delta = 0 \ . \tag{2.8-50}$$

Invariance of (2.8-48) imposes the restriction

$$f(0,0) = -\tau'(0) \ . \tag{2.8-51}$$

Hence

$$\lambda = -2\beta \ .$$

Thus (2.8-46,47,48,49) is invariant under the two-parameter (β,γ) group

$$\xi = \beta x + \gamma x t$$

$$\tau = 2\beta t + \gamma t^2 \qquad (2.8\text{-}52)$$

$$f = -2\beta - \gamma[\frac{x^2}{4} + \frac{t}{2}] \quad .$$

Corresponding to each of these parameters, we obtain the following similarity forms:

Case I $\gamma = 0$, $\beta = 1$ (stretching invariance).

The characteristic equations are:

$$\frac{dx}{x} = \frac{dt}{2t} = \frac{du}{-2u} \qquad (2.8\text{-}53)$$

The similarity variable is $\zeta_1 = \frac{x}{\sqrt{t}}$ and the similarity form for the solution is

$$u = \frac{1}{t} F_1(\zeta_1) \qquad (2.8\text{-}54)$$

where F_1 is an arbitrary function of ζ_1.

Case II $\gamma = 1$, $\beta = 0$.

In this case the characteristic equations are those of (2.8-30) with $x_o = 0$. The similarity variable is $\zeta_2 = \frac{x}{t}$ with the solution having the similarity form

$$u = \frac{e^{-\frac{x^2}{4t}}}{\sqrt{t}} F_2(\zeta_2) \quad . \qquad (2.8\text{-}55)$$

Equating the functional forms (2.8-54,55) and making trivial rearrangements, we see that

$$\frac{F_1(\zeta_1) e^{\frac{\zeta_1^2}{4}}}{\zeta_1} = \frac{F_2(\zeta_2)}{\zeta_2} = \text{constant} = C \quad . \qquad (2.8\text{-}56)$$

Hence

$$u = \frac{Cx}{t^{3/2}} e^{-\frac{x^2}{4t}} \quad . \tag{2.8-57}$$

The source condition implies that $C = \frac{1}{\sqrt{4\pi}}$,

$$\therefore \quad u = \frac{1}{\sqrt{4\pi}} \frac{x}{t^{3/2}} e^{-\frac{x^2}{4t}}$$

$$=> \qquad u(y,t) = \frac{1}{\sqrt{4\pi\nu}} \frac{y}{t^{3/2}} e^{-\frac{y^2}{4\nu t}} \quad . \tag{2.8-58}$$

The earthworm's Christmas problem

This problem is concerned with the propagation of damped temperature waves into the earth due to annual temperature variation. It illustrates the use of complex group parameters. We will approximate the annual temperature distribution at the earth's surface (taken to be the plane $x = 0$) by

$$u = T_1 + T_2 \cos At \tag{2.8-59}$$

where $A = \frac{2\pi}{\tau}$, $\tau = 1$ year, T_1 and T_2 are constants. Let $x > 0$ represent the distance from the earth's surface toward the centre. Then a reasonable approximate boundary condition to obtain the temperature distribution near the surface of the earth is to assume that $u(x \to \infty, t) = T_3 = \text{const.}$ The constants T_1, T_2 and T_3 are functions of longitude and latitude. Let $v = u - T_3$. Then the problem becomes:

$$\frac{\partial^2 v}{\partial x^2} = \frac{\partial v}{\partial t} \quad , \quad 0 < x < \infty \tag{2.8-60}$$

with boundary conditions

$$v(0,t) = (T_1 - T_3) + T_2 \cos At \quad , \tag{2.8-61}$$

$$v(\infty,t) = 0 \quad . \tag{2.8-62}$$

We set $v = v_1 + v_2$ where v_1 and v_2 both satisfy (2.8-60) and

$$\begin{aligned} v_1(0,t) &= T = T_1 - T_3 \\ v_2(0,t) &= T_2 \cos At \quad . \end{aligned} \tag{2.8-63}$$

$v_1(x,t)$ is the solution (2.5-30) obtained for the corresponding Rayleigh problem $(v_1(x,0) = 0)$. $v_2(x,0) = \Theta(x)$, the initial temperature distribution. However we are interested in finding the steady state solution and it can be shown that this solution is independent of the value of $\Theta(x)$. We now find the steady state solution $v_2(x,t)$ via similarity. Consider

$$\frac{\partial^2 V}{\partial x^2} = \frac{\partial V}{\partial t} \tag{2.8-64}$$

with boundary conditions

$$V(0,t) = T_2 e^{iAt} \tag{2.8-65}$$

$$V(x \to \infty,t) = 0 \quad . \tag{2.8-66}$$

The solution $V(x,t)$ to (2.8-64,65,66) will be a complex-valued function and $v_2(x,t) = \operatorname{Re} V(x,t)$.

Now we find the subgroup of the group (2.7-12) leaving invariant (2.8-65,66). Invariance of $x = 0$ implies that

$$\xi(0,t) = 0$$

$$\Rightarrow \qquad \kappa = \delta = 0 \quad .$$

Invariance of (2.8-65)

$$\Rightarrow \qquad f(0,t) = iA\tau(t) \quad . \tag{2.8-67}$$

Thus
$$-\frac{\gamma t}{2} + \lambda = iA(\alpha + 2\beta t + \gamma t^2)$$

Hence $\beta = \gamma = 0$ and $\lambda = iA\alpha$. The corresponding characteristic equations are

$$\frac{dx}{0} = \frac{dt}{1} = \frac{dV}{iAV} \quad .$$

The similarity variable is $\zeta = x$ and the solution has the similarity form

$$V = e^{iAt} F(x) \quad . \tag{2.8-68}$$

Substituting (2.8-68) into (2.8-64) we find that

$$F'' = iAF \quad , \tag{2.8-69}$$

Thus
$$F(x) = C_1 e^{x\sqrt{\frac{A}{2}}[1+i]} + C_2 e^{-x\sqrt{\frac{A}{2}}[1+i]} \tag{2.8-70}$$

where C_1 and C_2 are arbitrary constants.

$$F(\infty) = 0 \Rightarrow C_1 = 0$$

$$\Rightarrow \qquad C_2 = T_2 \quad .$$

Hence
$$V(x,t) = T_2 e^{i[At-\sqrt{\frac{A}{2}}\, x]} e^{-\sqrt{\frac{A}{2}}\, x} \tag{2.8-71}$$

$$\Rightarrow \quad v_2(x,t) = T_2 \cos\left(At - \sqrt{\frac{A}{2}}\, x\right) e^{-\sqrt{\frac{A}{2}}\, x} \quad , \tag{2.8-72}$$

If $t = \tau^*$ is an "earthperson's" Christmas (coldest time of the year, say), then an earthworm located at $x = x_o$ celebrates Christmas at time

$$t = \tau^* + \sqrt{\frac{1}{2A}}\, x_o \quad , \tag{2.8-73}$$

<u>An inverse Stefan problem</u>

As another example of an application of our similarity solutions to the heat equation we consider the problem of transient heat conduction in a melting slab[(4),(5)]. A one-dimensional finite slab originally extending from $x = 0$ to $x = x_o$ is melted in such a way that the face $x = 0$ is insulated[(6)] and the other face is melted with heat flowing into the melting face at a rate $H_o(t)$. It is assumed that all of the molten material is removed immediately upon formation. At time t the melting face is located at $x = X(t)$ with $X(0) = x_o$.

Let T_m be the melting temperature and $T_o(x)$ be the initial temperature distribution. Then the appropriate partial differential equation and boundary conditions are:

$$\begin{aligned}
&C\rho \frac{\partial T}{\partial t} = K \frac{\partial^2 T}{\partial x^2}, \quad 0 < x < X(t), \quad t > 0 \\
&T(X(t),t) = T_m, \quad t > 0 \\
&\frac{\partial T}{\partial x} = 0 \quad \text{at} \quad x = 0, \quad t > 0 \\
&T(x,0) = T_o(x), \quad 0 < x < x_o \\
&H_o(t) = +K \frac{\partial T}{\partial x} - \rho L \frac{dX}{dt} \quad \text{at} \quad x = X(t), \quad t > 0
\end{aligned} \tag{2.8-74}$$

where C = specific heat

ρ = density

K = thermal conductivity

L = latent heat of fusion.

We non-dimensionalize (2.8-74) by letting

---

(4) R. W. Sanders, "Transient Heat Conduction in a Melting Finite Slab: An Exact Solution", ARS Journal, November 1960, pp. 1030-1031.

(5) David Langford, "Pseudo-similarity Solutions of the One-dimensional Diffusion Equation with Applications to the Phase Change Problem", Quart. of Appl. Math., April 1967, pp. 45-52.

(6) The case where there is also a heat flux $H_1(t)$ at the face $x = 0$ will be considered at the end of this example.

$$y = \frac{x}{x_o}$$

$$\tau = \frac{Kt}{C\rho x_o^2}$$

$$Y(\tau) = \frac{X(t)}{x_o}$$

$$u(y,\tau) = \frac{T(x,t) - T_m}{T_m} \tag{2.8-75}$$

$$H(\tau) = \frac{cx_o}{LK} H_o(t)$$

$$k = \frac{cT_m}{L}$$

$$g(y) = \frac{T_o(x) - T_m}{T_m}$$

Then (2.8-74) becomes

$$\frac{\partial u}{\partial \tau} = \frac{\partial^2 u}{\partial y^2} , \quad 0 < y < Y(\tau) , \quad \tau > 0 \tag{2.8-76}$$

$$u(Y(\tau),\tau) = 0 , \quad \tau > 0 \tag{2.8-77}$$

$$\frac{\partial u}{\partial y} = 0 \quad \text{at} \quad y = 0 , \quad \tau > 0 \tag{2.8-78}$$

$$u(y,0) = g(y) , \quad 0 < y < 1 \tag{2.8-79}$$

$$H(\tau) = k \frac{\partial u}{\partial y} - \frac{dY}{d\tau} \quad \text{at} \quad y = Y(\tau) , \quad \tau > 0 \ . \tag{2.8-80}$$

Instead of seeking the direct solution of this nonlinear problem (we have to find both $X(t)$ and $T(x,t)$) we will look at the inverse Stefan problem, a "control" type of problem. In the inverse problem $Y(\tau)$ $(X(t))$ is specified and the solution is found which satisfies all of the boundary conditions except (2.8-80). The generated solution $u(y,\tau)$ fixes the value of $H(\tau)$. However it might be added that by appropriately piecing together "inverse" solutions it is possible to solve numerically the direct problem.[7]

In order to apply the "general" similarity solution of the heat equation to this problem we must leave invariant $y = 0$ and $y = Y(\tau)$ where $Y(0) = 1$.

We consider the four cases of invariant boundaries derived in §2.7.

Case I
$$\zeta = \frac{y - (A\tau + B)}{\sqrt{\alpha + 2\beta\tau + \gamma\tau^2}} .$$

In order to leave invariant $y = 0$, $y = Y(\tau)$ and have $Y(0) = 1$, we set $A = B = 0$ and $\alpha = 1$. Hence for this case we could solve the inverse Stefan problem for a two-parameter moving boundary

$$Y(\tau) = \sqrt{1 + 2\beta\tau + \gamma\tau^2}$$

with $\beta < 0$, $\gamma < 0$. Replacing $\beta \to -\beta$, $\gamma \to -\gamma$, the similarity variable is

$$\zeta = \frac{y}{\sqrt{1 - 2\beta\tau - \gamma\tau^2}} , \quad 0 \leq \zeta \leq 1 \tag{2.8-81}$$

where

$$\zeta = 0 \leftrightarrow y = 0$$
$$\zeta = 1 \leftrightarrow y = Y(\tau) .$$

The corresponding basic similarity solution is (cf. 2.7-18,19,19')

$$u(y,\tau) = \frac{F(\zeta)}{(1 - 2\beta\tau - \gamma\tau^2)^{1/4}} \left(\frac{C + \beta + \gamma\tau}{C - \beta - \gamma\tau} \right)^{\rho} e^{\frac{\gamma\tau\zeta^2}{4}} \tag{2.8-82}$$

where

$$F(\zeta) = G(z) e^{+\frac{\beta\zeta^2}{4}}$$

$$C = \sqrt{\beta^2 + \gamma}$$

---

(7) F. Milinazzo and G. W. Bluman, Numerical Similarity Solutions to Stefan Problems, unpublished manuscript.

$$z = \zeta\sqrt{C} \quad .$$

with $G(z)$ satisfying (2.7-19'), namely,

$$\frac{d^2G}{dz^2} + [\frac{1}{2} + \nu - \frac{1}{4} z^2]G = 0$$

where ν is arbitrary and

$$\rho = -\left[\frac{1}{2}\nu + \frac{1}{4}\right] .$$

(2.8-77,78) => $F'(0) = F(1) = 0$.

The subcase $\gamma = 0$ was considered by Sanders[8] following the work of Landau[9]. The moving boundary is eliminated by a "trick" change of variable. Then separation of variables is applied with ν playing the role of an eigenvalue in the superposition of similarity solutions of the basic form (2.8-82). Langford[10]'s solution also corresponds to $\gamma = 0$.

Case II

Here $$\zeta = [y + \delta + \frac{\kappa + \delta\beta}{2(\tau - \beta)}]\cdot\frac{1}{\tau - \beta} ,$$

replacing β by $-\beta$.

In order to leave invariant the boundaries $y = 0$ and $y = Y(\tau)$ and to have melting, we must set $\kappa = \delta = 0$, $\beta > 0$. The resulting melting boundary is

$$Y(\tau) = \frac{\beta - \tau}{\beta} = 1 - \frac{\tau}{\beta}$$

with similarity variable , after obvious relabelling,

$$\zeta = -\frac{\beta y}{\tau - \beta} , \quad 0 \le \zeta \le 1 \qquad (2.8\text{-}83)$$

---

(8) ibid.

(9) H. G. Landau "Heat Conduction in a Melting Solid", Quart. J. of Appl. Math., Vol. 8, No. 1, April 1950, pp. 81-94.

(10) ibid.

where
$$\zeta = 0 \leftrightarrow y = 0$$
$$\zeta = 1 \leftrightarrow y = Y(\tau) \quad .$$

The similarity curves are straight lines in (y,τ) space, filling a triangular region. When $\tau = \beta$, the bar has melted completely.

The basic similarity solution (cf. 2.7-20,21) corresponding to Case II is:

$$u(y,\tau) = \frac{F_\nu(\zeta)}{\sqrt{\beta - \tau}} e^{\frac{\nu^2\beta^2}{(\tau-\beta)}} e^{-\frac{\zeta^2}{4\beta^2}(\tau-\beta)} \tag{2.8-84}$$

with $F_\nu(\zeta)$ satisfying the simple differential equation

$$\frac{d^2F_\nu}{d\zeta^2} + \nu^2 F_\nu = 0 \tag{2.8-85}$$

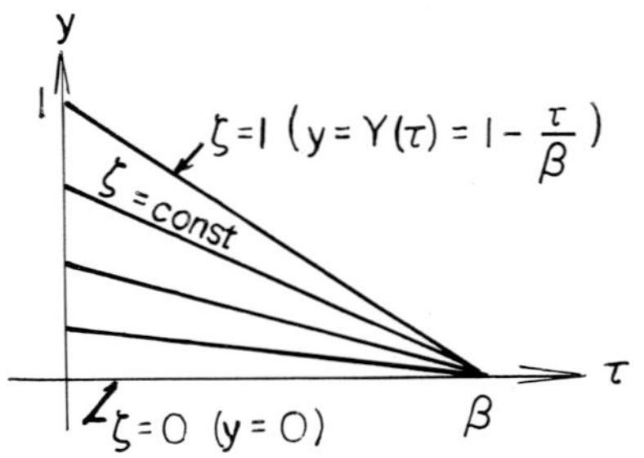

Figure 2.8-1

with ν arbitrary.

Cases III and IV are not amenable to melting problems.

Before proceeding any further, we will show that the solution to the inverse Stefan problem (2.8-76,77,78,79) is unique. Say $u = \theta_1$ and $u = \theta_2$ are solutions corresponding to $y = Y(\tau)$. Let $v = \theta_2 - \theta_1$. Then v satisfies

$$\frac{\partial v}{\partial \tau} = \frac{\partial^2 v}{\partial y^2}, \quad 0 < y < Y(\tau), \quad \tau > 0 \tag{2.8-76'}$$

$$v(Y(\tau),\tau) = 0, \quad \tau > 0 \tag{2.8-77'}$$

$$\frac{\partial v}{\partial y} = 0 \text{ at } y = 0, \quad \tau > 0 \tag{2.8-78'}$$

$$v(y,0) = 0, \quad 0 < y < 1 \quad . \tag{2.8-79'}$$

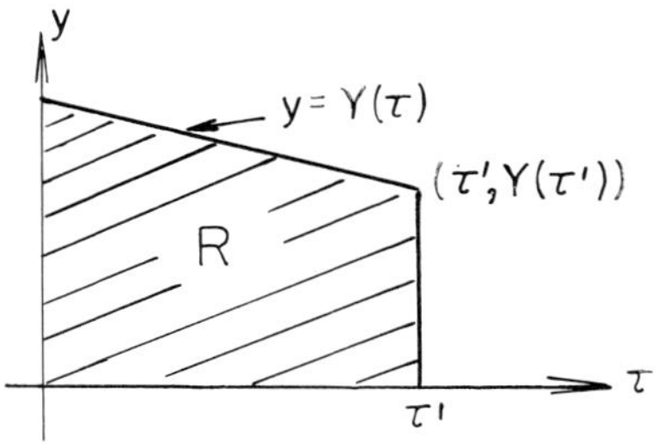

Figure 2.8-2

Since v satisfies (2.8-76') in region R, v also satisfies the identity

$$\frac{\partial}{\partial y}\left(v \frac{\partial v}{\partial y}\right) - \frac{1}{2} \frac{\partial v^2}{\partial \tau} = \left(\frac{\partial v}{\partial y}\right)^2$$

in R. Let C be the curve enclosing R. Then by Green's theorem

$$\oint_C \frac{1}{2} v^2 dy + v \frac{\partial v}{\partial y} d\tau = \iint_R \left(\frac{\partial v}{\partial y}\right)^2 dy d\tau$$

From the boundary conditions on v and $\frac{\partial v}{\partial y}$ along C we see that

$$\iint_R \left(\frac{\partial v}{\partial y}\right)^2 dy d\tau \equiv 0$$

$$\Rightarrow \qquad \frac{\partial v}{\partial y} \equiv 0 \quad \text{in} \quad R$$

$$\Rightarrow \qquad v = V(\tau) \quad \text{for some function of} \quad \tau \quad \text{in} \quad R .$$

But since v satisfies (2.8-76') we see that $V(\tau) = \text{constant} = V_o$. The boundary conditions imply that $V_o = 0$. Hence $\theta_1 = \theta_2$ and thus the solution to (2.8-76,77,78,79) is unique if $Y(\tau)$ is a specified melting curve.

We now consider in more detail Case II when the phase change boundary (melting face) moves at a constant velocity characterized by the parameter β.

$$(2.8\text{-}77,78) \Rightarrow$$

$$F_\nu(1) = 0 ,$$

$$F'_\nu(0) = 0 .$$

Hence

$$F_\nu(\zeta) = A \cos \nu\zeta \tag{2.8-86}$$

where ν takes on a discrete set of eigenvalues $\nu = \nu_n = \frac{(2n+1)}{2}\pi$, $n = 0,1,2,\ldots$.

Using superposition of these similarity solutions, we obtain the following solution satisfying (2.8-76,77,78):

$$u(y,\tau) = \sum_{n=0}^{\infty} \frac{A_n \cos \nu_n \zeta}{\sqrt{\beta - \tau}} e^{\frac{\nu_n^2 \beta^2}{\tau - \beta}} e^{-\frac{\zeta^2(\tau-\beta)}{4\beta^2}} \qquad (2.8\text{-}87)$$

We now use (2.8-79) to determine the arbitrary constants $\{A_n\}$.

$$u(y,0) = g(y) , \quad 0 < y < 1 \Rightarrow$$

$$g(y) = \sum_{n=0}^{\infty} \frac{A_n}{\sqrt{\beta}} \cos(n + \tfrac{1}{2})\pi y e^{-(n+\frac{1}{2})^2 \pi^2 \beta} e^{\frac{y^2}{4\beta}} . \qquad (2.8\text{-}88)$$

Let $\phi_n(y) = \cos(n + \frac{1}{2})\pi y$, $n = 0, 1, 2, \ldots$.

Then the eigenfunctions $\{\phi_n\}$ form a complete orthogonal set of functions on $[0;1]$ with

$$\int_0^1 \phi_n(y)\phi_m(y)dy = \frac{1}{2}\delta_{nm} .$$

Hence

$$A_n = 2\sqrt{\beta}\, e^{(n+\frac{1}{2})^2\pi^2\beta} \int_0^1 g(y)\phi_n(y) e^{-\frac{y^2}{4\beta}} dy , \qquad (2.8\text{-}89)^{(11)}$$

$n = 0, 1, 2, \ldots$.

<u>Problem 2.8-1:</u>

Show that if $g(y)$ is bounded, then the series solution (2.8-87) converge absolutely and uniformly in the region $0 \le \zeta \le 1$, $0 < \tau_o \le \tau \le \beta_o < \beta$, for arbitrary values of τ_o , β_o .

Now we consider a "reasonable" initial temperature distribution,

---

(11) If $\beta \to \infty$, i.e. the boundary $y = 1$ is fixed, then the solution (2.8-87,89) corresponds to the well-known Fourier series solution satisfying the boundary conditions (2.8-77,78,79) obtained by separation of variables.

namely

$$g(y) = -Be^{\frac{y^2}{4\beta}} \cos\frac{\pi}{2}y, \quad 0 < y < 1 \tag{2.8-90}$$

where $\beta > \frac{2}{\pi^2}$, $B < 1$.

In this case $A_o = B\sqrt{\beta}\, e^{\frac{\pi^2\beta}{4}}$ and $A_n = 0$, $n = 1,2,...$.
The resulting solution is

$$u(y,\tau) = -Be^{\frac{\pi^2\beta}{4}} \frac{\sqrt{\beta}}{\sqrt{\beta-\tau}} \cos\frac{\beta\pi y}{2(\tau-\beta)} e^{\frac{\pi^2\beta^2-y^2}{4(\tau-\beta)}},$$

$$0 \le y \le \frac{\beta-\tau}{\beta}, \tag{2.8-91}$$

$$0 \le \tau \le \beta .$$

At $y = Y(\tau) = 1 - \frac{\tau}{\beta}$,

$$\frac{\partial u}{\partial y} = \frac{B\pi}{2} e^{\frac{\pi^2}{4}\beta} \left(\frac{\beta}{\beta-\tau}\right)^{3/2} e^{\frac{\pi^2\beta^4-(\beta-\tau)^2}{4\beta^2(\tau-\beta)}} .$$

Hence

$$H(\tau) = \frac{kB\pi}{2} e^{\frac{\pi^2\beta}{4}} \left(\frac{\beta}{\beta-\tau}\right)^{3/2} e^{\frac{\pi^2\beta^4-(\beta-\tau)^2}{4\beta^2(\tau-\beta)}} + \frac{1}{\beta} . \tag{2.8-92}$$

<u>Problem 2.8-2</u>:

Show that for <u>any</u> value of β , $H(\tau)$ is a monotonically decreasing function of τ $(0 \le \tau < \beta)$. Thus

$$H(\beta) = \frac{1}{\beta} \le H(\tau) \le \frac{1}{\beta} + \frac{kB\pi}{2} e^{\frac{1}{4\beta}} = H(0)$$

<u>Problem 2.8-3</u>:

Construct the similarity solution corresponding to Case I.

If there is a (nondimensionalized) heat flux $h(\tau)$ at the fixed end $y = 0$, then the corresponding boundary value problem is:

$$\frac{\partial u}{\partial t} = \frac{\partial^2 u}{\partial y^2} \quad , \quad 0 < y < Y(\tau) \quad , \quad \tau > 0 \tag{2.8-93}$$

$$u(Y(\tau),\tau) = 0 \quad , \quad \tau > 0 \tag{2.8-94}$$

$$\frac{\partial u}{\partial y}(0,\tau) = h(\tau) \quad , \quad \tau > 0 \tag{2.8-95}$$

$$u(y,0) = g(y) \quad , \quad 0 < y < 1 \tag{2.8-96}$$

$$H(\tau) = k\,\frac{\partial u}{\partial y}(Y(\tau),\tau) - \frac{dY}{d\tau} \quad , \quad \tau > 0 \quad . \tag{2.8-97}$$

We assume Case II, i.e., $Y(\tau)$ is of the form

$$Y(\tau) = 1 - \frac{\tau}{\beta} \quad . \tag{2.8-98}$$

In order to leave invariant $y = 0$ and $y = Y(\tau)$, the similarity variable is

$$\zeta = +\frac{\beta y}{\beta - \tau} \quad , \quad 0 \le \zeta \le 1 \quad . \tag{2.8-99}$$

The corresponding similarity solution is (2.8-84) with $F_\nu(\zeta)$ satisfying (2.8-85). Uniqueness of the solution to this inverse Stefan problem follows from the same argument as for the case when $h(\tau) = 0$. Invariance of (2.8-94)

$$=> \qquad F_\nu(1) = 0 \tag{2.8-100}$$

Hence

$$F_\nu(\zeta) = A_\nu \sin \nu(\zeta - 1) \quad , \tag{2.8-101}$$

A_ν arbitrary.

Thus an arbitrary formal linear superposition:

$$u(y,\tau) = \sum_\nu \frac{A_\nu \sin \nu(\zeta - 1)}{\sqrt{\beta - \tau}}\, e^{\frac{\nu^2\beta^2}{\tau - \beta}}\, e^{-\frac{\zeta^2(\tau - \beta)}{4\beta^2}} \tag{2.8-102}$$

is a solution to (2.8-93,94), $0 < \tau < \beta$, and moreover $\zeta = 0 \leftrightarrow y = 0$. In order to satisfy (2.8-95) we first set

$$r = \frac{\beta\tau}{\beta - \tau}$$

$$s = -\nu^2 \tag{2.8-103}$$

$$B(s) = \frac{A_\nu e^{-\nu^2\beta}}{\beta}$$

and formally replace $\sum_\nu$ by $\frac{1}{2\pi i}\int_{\gamma-i\infty}^{\gamma+i\infty} ds$

=> the following inverse Laplace transform with unknown $B(s)$ yields a solution

$$u = \theta_1(y,\tau) = \sqrt{r+\beta}\; e^{\frac{\zeta^2}{4(r+\beta)}} \cdot \frac{1}{2\pi i}\int_{\gamma-i\infty}^{\gamma+i\infty} B(s)[\sinh\sqrt{s}(1-\zeta)]e^{sr}ds\,. \tag{2.8-104}$$

Letting

$$h(r) = h(\tau) = h\left(\frac{\beta r}{r+\beta}\right) \tag{2.8-105}$$

and inverting (2.8-104) such that the boundary condition (2.8-95) is satisfied

$$=> \qquad B(s) = \frac{-\beta}{\sqrt{s}\cosh\sqrt{s}}\int_0^\infty \frac{h(r)}{(r+\beta)^{3/2}} e^{-sr}dr\,. \tag{2.8-106}$$

The corresponding solution $u = \theta_1(y,\tau)$ now satisfies (2.8-93,94,95).

Next we compute $\theta_1(y,0) = g^*(y)$. In order to find a solution

$$u = \theta_1(y,\tau) + \theta_2(y,\tau) \qquad \text{satisfying}$$

(2.8-93,94,95,96,97), we find the solution $\theta_2(y,\tau)$ of the heat equation satisfying the boundary conditions for zero heat flux at $y = 0$, namely, (2.8-77,78,79, $Y(\tau) = 1 - \frac{\tau}{\beta}$) with $g(y)$ replaced by $g(y) - g^*(y)$. The generated heat flux will be

$$H(\tau) = k\left[\frac{\partial\theta_1}{\partial y}(Y(\tau),\tau) + \frac{\partial\theta_2}{\partial y}(Y(\tau),\tau))\right] + \frac{1}{\beta}\,. \tag{2.8-107}$$

Burgers' problem

A special case of the solutions (2.7-22,23) has been used by

J. M. Burgers[12] to represent eigensolutions of the heat equation for a semi-infinite space bounded by a moving boundary of parabolic form

$$x = \frac{1}{2} t^2 \tag{2.8-108}$$

(see Fig. 2.8-3). These eigensolutions are obtained from (2.7-22,23) if

$$\kappa = 0 \ , \ \delta = 1 \ , \quad \zeta^* = -\zeta \tag{2.8-109}$$

and

$$\lambda = -\sigma_m \tag{2.8-110}$$

plays the role of an eigenvalue. Equation (2.7-22) becomes

$$u_m(x,t) = F_m(\zeta^*)\exp[-\frac{1}{12} t^3 + \frac{1}{2} \zeta^* t - \sigma_m t] \ ,$$

$$\zeta^* = -\zeta = \frac{1}{2} t^2 - x \ . \tag{2.8-111}$$

The $\{F_m(\zeta^*)\}$ are the Airy function solutions of (2.7-23) which die out as $\zeta^* \to \infty$,

$$\frac{d^2F_m}{d\zeta^{*2}} - (\frac{1}{2} \zeta^* - \sigma_m)F_m = 0 \ , \tag{2.8-112}$$

$$F_m(\zeta^*) = \frac{(\zeta^* - 2\sigma_m)^{1/2}}{2^{1/6}3^{1/3}} K_{\frac{1}{3}} \left(\frac{2^{1/2}}{3} [\zeta^* - 2\sigma_m]^{3/2} \right) . \tag{2.8-113}$$

The σ_m are chosen so that $(-2\sigma_m)$ $m = 1, 2, \ldots,$ are the zeroes of the Airy function or of the continuation of (2.8-113) to negative arguments

$$F_m(\zeta^*) = \frac{\pi(2\sigma_m - \zeta^*)^{1/2}}{2^{1/6}3^{5/6}} \cdot \left\{ J_{-\frac{1}{3}} \left(\frac{2^{1/2}}{3}[2\sigma_m - \zeta^*]^{3/2} \right) + J_{\frac{1}{3}} \left(\frac{2^{1/2}}{3}[2\sigma_m - \zeta^*]^{3/2} \right) \right\} . \tag{2.8-114}$$

Thus $F_m(0) = 0 \ \forall \ m$. Burgers shows that this set of eigenfunctions

---

(12) J. M. Burgers "Functions and Integrals Connected with Solutions of the Diffusion or Heat Flow Equation", University of Maryland, Technical Note, BN-398, May, 1965.

is orthogonal. He then uses them to construct a representation of a unit source at (x_s, t_s) reflected in the moving boundary. This representation is

$$u(x,t;x_s,t_s) =$$

$$\frac{1}{2}\exp[-\frac{1}{12}(t^3-t_s^3) + \frac{1}{2}(t\zeta^* - t_s\zeta_s^*)] \cdot \sum_{m=0}^{\infty} \frac{F_m(\zeta^*)F_m(\zeta_s^*)}{[F_m^*(0)]^2} e^{-\sigma_m(t-t_s)} ,$$

$$t \geq t_s \quad ; \quad u = 0 \quad , \quad t \leq t_s \quad . \qquad (2.8\text{-}115)$$

Burgers' method of deriving (2.8-111) is to assume a separation of variables form $u = G(t,\zeta^*)H(t,\zeta^*)$ and to choose G so that essentially a similarity form results for H .

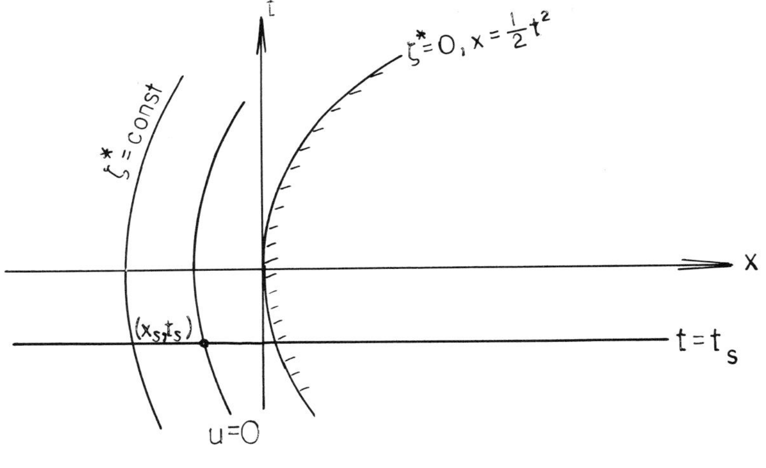

Figure 2.8-3

<u>Summary</u>.

In applying group invariance to boundary value problems for the heat equation the following points have been illustrated:

(i) Invariance under a multi (>1)-parameter group leads to a further reduction in the number of variables without use of the governing partial differential equation.

(ii) The group parameters may be complex.

(iii) If not all of the boundary conditions are invariant then a superposition of similarity solutions may be possible. This leads to the possibility of numerical techniques based on similarity, in particular for direct Stefan problems of the heat equation.

2.9.-Axially-Symmetric Wave Equation

The axially symmetric wave equation has sufficient structure to provide an interesting example in the application of our methods. In this section a (fairly) general group is found and then application is made to the problem of the fundamental solution. This particular problem is invariant under a two-parameter group so that the solution can be found from inspection of a functional equation (cf §2.8).

The basic equation considered, at first, is

$$Lu \equiv u_{tt} - u_{rr} - \frac{1}{r} u_r = 0 \ , \quad r = \sqrt{x^2 + y^2} \quad . \tag{2.9-1}$$ [1]

Initial and boundary conditions are put aside until later. In searching for the group of transformations which leaves invariant (2.9-1) it is sufficiently general to consider the form

$$\begin{aligned} t^* &= t + \varepsilon\tau(r,t) \\ r^* &= r + \varepsilon\rho(r,t) \\ u^* &= u + \varepsilon u g(r,t) \quad . \end{aligned} \tag{2.9-2}$$ [2]

Then, using the formulas for the first and second extensions, as worked out in §2.1 (2.1-19,21,22), it is easy to write down the equation satisfied by $u^* = \Theta^*(x^*,t^*) = U(x,t,\Theta(x,t);\varepsilon)$ when $u = \Theta(x,t)$:

---

(1) Units are chosen to make the wave speed one.

(2) cf. Ovsjannikov, Chap. 6, ibid.

$$L^*\Theta^* = L\Theta + \varepsilon\{\Theta Lg + \Theta_t(2g_t - L\tau) + \Theta_r(-L\rho + \frac{\rho}{r^2} - 2g_r - \frac{g}{r}) + \Theta_{tt}(g - 2\tau_t) + \Theta_{rt}(-2\rho_t + \tau_r) + \Theta_{rr}(-g + 2\rho_r)\} \quad . \tag{2.9-3}$$

Then, for $L\Theta = 0$, we use $\Theta_{tt} = \Theta_{rr} + \frac{1}{r}\Theta_r$ and

$$L^*\Theta^* = \varepsilon\{\Theta Lg + \Theta_t(2g_t - L\tau) + \Theta_r(-L\rho + \frac{\rho}{r^2} - 2g_r - \frac{2}{r}\tau_t) + \Theta_{rr}(-2\rho_r - 2\tau_t) + \Theta_{rt}(-2\rho_t + 2\tau_r)\} . \tag{2.9-4}$$

The condition of invariance is that the $0(\varepsilon)$ terms in (2.9-4) vanish. Since Θ, Θ_t, Θ_r, Θ_{rr}, Θ_{rt} are independent each coefficient must vanish and the following conditions on the infinitesimals (τ, ρ, g) arise:

$$Lg = g_{tt} - g_{rr} - \frac{1}{r}g_r = 0 \tag{2.9-5}$$

$$2g_t - L\tau = 0 \tag{2.9-6}$$

$$- L\rho + \frac{\rho}{r^2} - 2g_r - \frac{2}{r}\tau_t = 0 \tag{2.9-7}$$

$$\rho_r - \tau_t = 0 \tag{2.9-8}$$

$$\rho_t - \tau_r = 0 \quad . \tag{2.9-9}$$

The last two imply

$$\rho_{tt} - \rho_{rr} = 0 \ , \ \tau_{tt} - \tau_{rr} = 0 \tag{2.9-10}$$

Then, from (2.9-6) and (2.9-9) we deduce

$$g = - \frac{\rho}{2r} + A(r)$$

and from (2.9-7)

$$g = - \frac{\rho}{2r} + B(t)$$

so that

$$g = -\frac{1}{2}\frac{\rho(r,t)}{r} + C \ , \quad C = \text{const.} \tag{2.9-11}$$

Then, (2.9-5) becomes

$$\frac{\rho_r}{\rho} = \frac{1}{r} \tag{2.9-12}$$

and

$$\rho = D(t)r \quad . \tag{2.9-13}$$

Since $\rho_{tt} - \rho_{rr} = 0$, $D(t) = \alpha + \beta t$ and then it is easy to solve for τ from (2.9-8,9). The final results for the infinitesimals of the group are

$$\rho(r,t) = \alpha r + \beta r t \tag{2.9-14}$$

$$\tau(r,t) = \gamma + \alpha t + \frac{1}{2}\beta(r^2 + t^2) \tag{2.9-15}$$

$$g(r,t) = \delta - \frac{1}{2}\beta t \quad . \tag{2.9-16}$$

The similarity variables, and the functional form of the solution all follow from the invariant surface condition (2.1-8 ff.). The corresponding characteristic equations are:

$$\frac{dt}{\gamma + \alpha t + \frac{1}{2}\beta(r^2 + t^2)} = \frac{dr}{\alpha r + \beta r t} = \frac{du}{(\delta - \frac{1}{2}\beta t)u} \quad . \tag{2.9-17}$$

Three of the four (α, β, γ, δ) are thus independent. As far as (r,t) are concerned, γ by itself corresponds to translation in t , α by itself to the stretching group.

<u>Problem 2.9-1</u>: Work out the commutator table corresponding to (2.9-14,15,16).

Some special similarity forms, which are useful for the problem of the fundamental solution are now worked out.

(i) $\gamma = \beta = 0$

$$\frac{dt}{\alpha t} = \frac{dr}{\alpha r} = \frac{du}{\delta u} \tag{2.9-18}$$

The integral of the first two yields the similarity variable

$$\zeta_1 = \frac{r}{t} \tag{2.9-19}$$

and integrating $\frac{dt}{\alpha t} = \frac{du}{\delta u}$ along $\zeta_1 = \text{const.}$ gives

$$\frac{\delta}{\alpha} \log t = \log u - \log F_1(\zeta_1)$$

or

$$u(r,t) = t^{\kappa} F_1(\zeta_1) \ ; \ \kappa = \frac{\delta}{\alpha} \quad . \tag{2.9-20}$$

(ii) $\gamma = \alpha = 0$

$$\frac{dt}{\frac{1}{2}\beta(r^2 + t^2)} = \frac{dr}{\beta rt} = \frac{du}{(\delta - \frac{1}{2}\beta t)u} \tag{2.9-21}$$

The first two yield

$$\frac{dt}{dr} = \frac{1}{2}\left(\frac{r}{t} + \frac{t}{r}\right) \quad . \tag{2.9-22}$$

Because of stretching invariance, the integrating factor is easily found, or the equation for s is easily solved if $t = rs(r)$. The result is

$$\zeta_2 = \frac{t^2}{r} - r \quad . \tag{2.9-23}$$

The functional form is obtained by integrating the last two of (2.9-21) with t replaced by $\sqrt{r^2 + r\zeta_2}$

$$\frac{du}{u} = \frac{\nu}{r\sqrt{r^2 + r\zeta_2}} dr - \frac{1}{2}\frac{dr}{r} \quad , \ \nu = \frac{\delta}{\beta} \quad , \tag{2.9-24}$$

Integration yields

$$u(r,t) = \frac{e^{-\frac{2\nu t}{t^2 - r^2}}}{\sqrt{r}} F_2(\zeta_2) \quad , \quad \zeta_2 = \frac{t^2}{r} - r \quad . \tag{2.9-25}$$

Fundamental solutions

The fundamental solution $u = S(r,t)$ is defined as the response to a concentrated unit-"impulse":

$$LS = S_{tt} - S_{rr} - \frac{1}{r} S_r = \frac{\delta(r)}{2\pi r}\delta(t) \tag{2.9-26}$$

with quiescent initial conditions

$$S(r,0-) = S_t(r,0-) = 0 \quad . \tag{2.9-27}$$

The solution is zero outside the cone of influence.

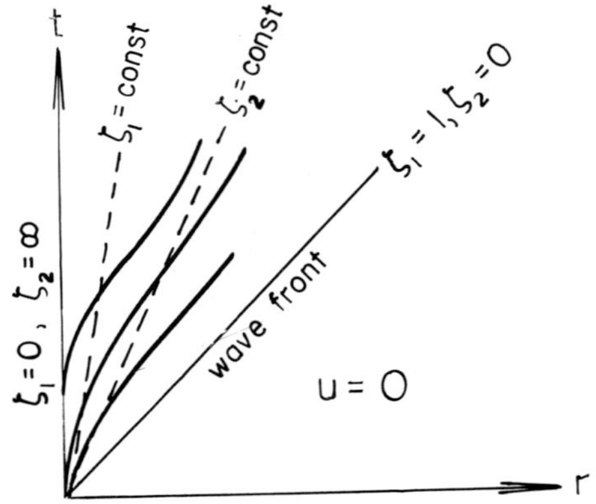

Figure 2.9-1

In order for a similarity representation to be useful for this class of problems, it is necessary that the wave-front $t - r = 0$ be left invariant. But if $\gamma = 0$ in (2.9-14,15), then $\tau - \rho = 0$

$$\Rightarrow \qquad t^* - r^* = 0 \quad \text{when} \quad t - r = 0$$

so that both (i) and (ii) above meet the requirements. In order to determine the appropriate value of the parameter in (2.9-20) we could study the local transformation of the δ-functions. But it is just as easy to use the global stretching transformation:

$$\begin{aligned} r^* &= \lambda r \\ t^* &= \lambda t \\ u^* &= \lambda^{\kappa} u \end{aligned} \tag{2.9-28}$$

which corresponds to (i). Thus, since

$$L^*S^* = \lambda^{\kappa-2}LS = \lambda^{\kappa-2}\frac{\delta\left(\frac{r^*}{\lambda}\right)\delta\left(\frac{t^*}{\lambda}\right)}{r^*/\lambda} = \lambda^{\kappa+1}\frac{\delta(r^*)\delta(t^*)}{r^*} \qquad (2.9\text{-}29)$$

the initial conditions at $t = 0$ are evidently invariant. Therefore, for invariance of the whole problem $\kappa = -1$ and

$$u = S(r,t) = \frac{1}{t}F_1(\zeta_1), \quad \zeta_1 = \frac{r}{t} \quad . \qquad (2.9\text{-}30)$$

It is also possible to deduce something about the forms of (2.9-25) but here we rely on uniqueness and set

$$u = S(r,t) = \frac{1}{t}F_1(\zeta_1) = \frac{e^{-\frac{2\nu t}{t^2-r^2}}}{\sqrt{r}}F_2(\zeta_2) \quad , \qquad (2.9\text{-}31)$$

Using

$$t = \frac{\zeta_1\zeta_2}{1-\zeta_1^2} \quad , \quad r = \frac{\zeta_1^2\zeta_2}{1-\zeta_1^2} \qquad (2.9\text{-}32)$$

we have

$$\frac{1-\zeta_1^2}{\zeta_1\zeta_2}F_1(\zeta_1) = \frac{e^{-\frac{2\nu}{\zeta_1\zeta_2}}}{\zeta_1\sqrt{\zeta_2}}\sqrt{1-\zeta_1^2}\,F_2(\zeta_2) \quad \text{for all } \zeta_{1,2} \ . \qquad (2.9\text{-}33)$$

This is only possible if the variables can be separated, so that $\nu = 0$ and

$$\sqrt{1-\zeta_1^2}\,F_1(\zeta_1) = \sqrt{\zeta_2}\,F_2(\zeta_2) = k = \text{const.} \qquad (2.9\text{-}34)$$

Thus the functional form of the fundamental solution is

$$u = S(r,t) = \frac{k}{\sqrt{t^2-r^2}} \quad . \qquad (2.9\text{-}35)$$

The evaluation of the constant k, does not follow directly from symmetry but depends on the overall conservation law corresponding to (2.9-26):

$$\int_{0-}^{t} dt' \int_0^\infty r'dr'(S_{t't'} - S_{r'r'} - \frac{1}{r'}S_{r'}) = \frac{1}{2\pi} \quad . \qquad (2.9\text{-}36)$$

Due to the singularity of (2.9-35) at the wave front $t = r$, the integrals in (2.9-36) have to be carried out carefully, say up to $t' = r' + \delta$, and then $\delta \to 0$ (cf Fig. 2.9-2). We note

$$S_t = -k\frac{t}{(t^2 - r^2)^{3/2}}, \quad S_r = k\frac{r}{(t^2 - r^2)^{3/2}} \tag{2.9-37}$$

then

$$\begin{aligned} I_1 &= \int_0^{t-\delta} r'dr' \int_{r'+\delta}^{t} S_{t't'}dt' \\ &= \int_0^{t-\delta} r'dr'\{S_t(r',t) - S_t(r',r'+\delta)\} \qquad (2.9\text{-}38) \\ &= \int_0^{t-\delta} r'dr'\left\{-k\frac{t}{(t^2 - r'^2)^{3/2}} + k\frac{r' + \delta}{(2\delta r' + \delta^2)^{3/2}}\right\} \\ &= k - k\frac{t}{\sqrt{2\delta t - \delta^2}} + \frac{k}{(2\delta)^{3/2}}\int_0^{t-\delta}\frac{r'^2 + \delta r'}{(r' + \delta/2)^{3/2}}dr'. \end{aligned}$$

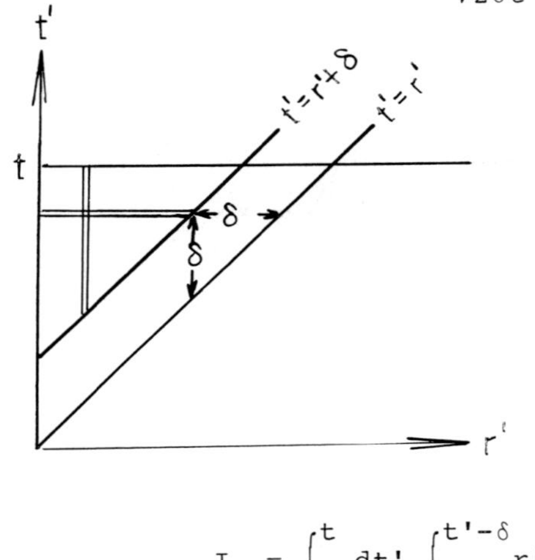

Figure 2.9-2

$$\begin{aligned} I_2 &= \int_\delta^t dt' \int_0^{t'-\delta} r'dr'(S_{r'r'} + \frac{1}{r'}S_{r'}) \\ &= \int_\delta^t (r'S_{r'})\Big|_0^{t'-\delta} dt' \\ &= k\int_\delta^t dt' \frac{(t' - \delta)^2}{[(t')^2 - (t' - \delta)^2]^{3/2}}, \quad \text{let } t' = \delta + r', \end{aligned}$$

$$\Rightarrow \quad I_2 = \frac{k}{(2\delta)^{3/2}}\int_0^{t-\delta}\frac{r'^2dr'}{(r' + \delta/2)^{3/2}} \quad . \tag{2.9-39}$$

Thus, (2.9-26) becomes

$$I_1 - I_2 = k - k\frac{\sqrt{t}}{\sqrt{2\delta}} + 0(\sqrt{\delta}) + \frac{k}{2^{3/2}\sqrt{\delta}}\int_0^{t-\delta}\frac{r'dr'}{(r' + \delta/2)^{3/2}}$$

$$I_1 - I_2 = k + 0(\sqrt{\delta}) \rightarrow k = \frac{1}{2\pi} \quad . \tag{2.9-40}$$

The final result is, from (2.9-35)

$$S(r,t) = \frac{1}{2\pi\sqrt{t^2 - r^2}} \quad . \tag{2.9-41}$$

Problem 2.9-2

Apply the method of this section to the one-dimensional wave equation

$$u_{tt} - u_{xx} = 0$$

for the class of problems with $u(x,0) = 0$. Show that the general similarity form is

$$u(x,t) = F(\zeta) \qquad \text{where} \qquad \zeta = \Omega(x + t) - \Omega(x - t) \ ,$$

Ω = an arbitrary function. Find $F(\zeta)$ from the wave equation.

Problem 2.9-3

Apply the method of this section to find the form of the fundamental solution for the telegraph equation

$$u_{tt} - u_{xx} + k^2u = \delta(x)\delta(t)$$

$$u(x,0-) = u_t(x,0-) = 0 \quad .$$

Show that $u = S(x,t) = F(\zeta)$, $\zeta = \sqrt{t^2 - x^2}$ and determine F . Derive the global transformation (Lorentz) corresponding to the infinitesimals found.

Problem 2.9-4

Use the methods of §2.8 to show that the subgroup of (2.9-14, 15,16) leaving invariant (2.9-26) corresponds to $\gamma = 0$, $\delta = -\alpha$. Hence show that the results obtained in this section can be derived in an easier fashion by taking each of the two remaining parameters in turn to generate similarity forms for the solution.

Problem 2.9-5

Find the fundamental solution (Riemann function) of the Euler-Poisson-Darboux equation, i.e. solve

$$u_{xx} - u_{yy} + \frac{\lambda}{x} u_x = \delta(x - x_o)\delta(y - y_o) , \qquad (2.9\text{-}42)$$

This solution is the source solution for isentropic flow for a polytropic gas.

x is the sound speed in the gas ,

y is the fluid velocity in some fixed direction ,

u is the time variable ,

λ is related to the ratio of specific heats of the gas.

Show that

(i) if

$$\begin{aligned} x^* &= x + \varepsilon\xi(x,y) + 0(\varepsilon^2) \\ y^* &= y + \varepsilon\eta(x,y) + 0(\varepsilon^2) \\ u^* &= u + \varepsilon g(x,y)u + 0(\varepsilon^2) \end{aligned} \qquad (2.9\text{-}43)$$

leaves invariant (2.9-42) then

$$\xi = -\frac{2x}{\lambda}(y - y_o)$$
$$\eta = \frac{x_o^2 - x^2 - (y - y_o)^2}{\lambda} \qquad (2.9\text{-}44)$$
$$g = y - y_o \quad .$$

(ii) the similarity variable corresponding to (2.9-44) is

$$\zeta = \frac{(x - x_o)^2 - (y - y_o)^2}{x}$$

and derive the similarity form of the solution.

(iii) the solution is

$$u(x,y) = \frac{(2x_o)^\lambda}{[(x + x_o)^2 - (y - y_o)^2]^{\lambda/2}} F\left(\frac{\lambda}{2}, \frac{\lambda}{2}, 1; \frac{(x-x_o)^2 - (y-y_o)^2}{(x+x_o)^2 - (y-y_o)^2} \right)$$

Note that the solution for a ring source in an axially-symmetric wave equation is contained in this formula. ($\lambda = 1$)

Problem 2.9-6

Consider the equation for the response of a string with a non-linear restoring force to a unit impulse:

$$u_{tt} - u_{xx} + f(u) = \delta(x)\delta(t)$$

$$u(x,0-) = u_t(x,0-) = 0 \ , \quad -\infty < x < \infty$$

$$f(u) = -f(-u) \ ; \quad f(u) > 0 \ , \quad u > 0 \quad ,$$

By applying the methods of this section find the similarity form of the solution. For the case $f(u) = \kappa u^3$ study the solution in a suitable phase plane. What conditions apply at the wave-front?

2.10. Similarity Solutions of the One-Dimensional Fokker-Planck Equation[1]

In this section we will illustrate a more complex example of invariance under a multi-parameter group by finding probability distributions which are fundamental solutions of the Fokker-Planck equation. It will be shown that moments are easily computable from invariance considerations and in turn new integrals involving special functions are derived since $\int_R p\,dx = 1$ for a probability distribution p on region R.

We consider the stochastic differential equation

$$\frac{dx}{dt} + f(x) = n(t) \tag{2.10-1}$$

where $n(t)$ represents stationary Gaussian white noise, in particular

$$\left.\begin{aligned} &\langle n(t)\rangle = 0 \\ &\langle n(t_1)n(t_2)\rangle = D\delta(t_2-t_1) \quad \text{for some constant } D \\ \text{and}\quad &x(0) = x_o. \end{aligned}\right\} \quad (2.10\text{-}2)^{(2)}$$

x can be thought of as the velocity of some particle influenced by a frictional force $f(x)$. The problem of a free particle in Brownian motion with Stokes' drag corresponds to the case $f(x) = \beta x$ for some constant β.

The output $x(t)$ of (2.10-1) is a stationary Markov process and is completely specified by finding the transitional probability density distribution $p(x,t|x_o) \geq 0$ which satisfies the Fokker-Planck equation

---

(1) G. W. Bluman, Similarity solutions of the one-dimensional Fokker-Planck equation, Int. J. Non-Lin. Mech., 6, pp. 143-153, 1971.

(2) $\langle n(t)\rangle$ denotes the expected value of $n(t)$.

$$\left.\begin{aligned} &\frac{\partial p}{\partial t} = D \frac{\partial^2 p}{\partial x^2} + \frac{\partial}{\partial x}[f(x)p] \\ &\text{with the initial condition} \\ &p(x,0|x_o) = \delta(x-x_o). \end{aligned}\right\} \quad (2.10\text{-}3)$$

$p(x,t|x_o)dx$ gives the probability that if initially a particle has velocity x_o then at time t its velocity lies in the region $[x; x+dx]$. $x = r$ is said to be a reflecting boundary for process (2.10-1,2) if

$$\lim_{x \to r} [D \frac{\partial p}{\partial x} + f(x)p] = 0. \quad (2.10\text{-}4)$$

If r_1 and r_2 are reflecting boundaries, $r_1 < r_2$, and $x_o \in R = (r_1; r_2)$, then

$$\int_{r_1}^{r_2} p(x,t|x_o)dx = 1. \quad (2.10\text{-}5)^{(3)}$$

The probability distribution $p(x,t|x_o)$ is said to be stable if

$$\lim_{t \to \infty} \int_{r_1}^{r_2} x^2 p(x,t|x_o)dx < \infty. \quad (2.10\text{-}6)$$

We limit our discussion to those processes for which $f(x)$ is an odd function, i.e.,

$$f(x) = -f(-x). \quad (2.10\text{-}7)$$

Our aim is to find the most general $f(x)$ leading to analytical solutions of (2.10-3) via similarity.

(3) Note that the heat equation corresponds to the Fokker-Planck equation for which $f(x) = 0$. Moreover, for the heat equation in §2.8 we found various fundamental solutions corresponding to the temperature distributions over (i) an infinite bar $(R = (-\infty; \infty))$, (ii) an insulated semi-infinite bar $(R = (0; \infty))$, and (iii) a finite bar insulated at both ends $(R = (r_1; r_2), -\infty < r_1 < r_2 < \infty)$.

In addition to the case of a free particle in Brownian motion, closed form solutions have been found for the cases $f(x) = k \operatorname{sgn} x$, $R = (-\infty; \infty)$, by Caughey and Dienes[4] and $f(x)$ piecewise linear by Atkinson and Caughey.[5]

By a simple transformation it is easy to see that (2.10-3) is equivalent to the system

$$\frac{\partial^2 p}{\partial x^2} + \frac{\partial}{\partial x}[f(x)p] = \frac{\partial p}{\partial t}$$
$$p(x,0|x_o) = \delta(x-x_o). \tag{2.10-3'}$$

Say the one-parameter Lie group

$$\begin{aligned} x^* &= X^*(x,t,p;\ \varepsilon) = x + \varepsilon\xi(x,t,p) + \cdots \\ t^* &= T^*(x,t,p;\ \varepsilon) = t + \varepsilon\tau(x,t,p) + \cdots \\ p^* &= P^*(x,t,p;\ \varepsilon) = p + \varepsilon\pi(x,t,p) + \cdots \end{aligned} \tag{2.10-8}$$

leaves invariant (2.10-3'). It turns out that ξ and τ do not depend on p and moreover $\pi(x,t,p)$ has the form

$$\pi(x,t,p) = g(x,t)p. \tag{2.10-9}$$

Corresponding to the one-parameter Lie group (2.10-8) we generate an infinitesimal operator

$$\mathfrak{X} = \xi(x,t)\frac{\partial}{\partial x} + \tau(x,t)\frac{\partial}{\partial t} + g(x,t)p\frac{\partial}{\partial p}. \tag{2.10-10}$$

The solution $p = P(x,t)$ of (2.10-3') satisfies the invariant surface condition

$$\mathfrak{X}\{p - P(x,t)\} = 0. \tag{2.10-11}$$

---

(4) T. K. Caughey and J. K. Dienes, The behaviour of linear systems with random parametric excitation, J. Math. Phys. 41, 300(1962).

(5) J. D. Atkinson and T. K. Caughey, Spectral density of piecewise linear first order systems excited by white noise, Int. J. Non-Lin. Mech, 3, 137(1968).

We recall that the resulting characteristic equations corresponding to this first order partial differential equation are

$$\frac{dx}{\xi(x,t)} = \frac{dt}{\tau(x,t)} = \frac{dp}{g(x,t)p} \, . \tag{2.10-12}$$

The integral of the first equality in (2.10-12) yields the similarity variable

$$\zeta(x,t) = \text{const.} \tag{2.10-13}$$

Then substituting this equation into the second equality of (2.10-12) we are led to the similarity form

$$p = F(\zeta)G(x,t) \, . \tag{2.10-14}$$

The dependence of G on x and t is known explicitly and $F(\zeta)$ is some arbitrary function of ζ. If the infinitesimal operator corresponding to the group (2.10-8) contains only one-parameter then as mentioned previously we substitute (2.10-14) into (2.10-3') to obtain a linear second order ordinary differential equation for $F(\zeta)$.

We now further discuss the case where a two-parameter (a,b) Lie group leaves invariant a partial differential equation. We previously encountered this situation in §2.8 (temperature distribution over an infinite bar) and in §2.9 (fundamental solution of the wave equation). Let $\mathfrak{X}_1$ and $\mathfrak{X}_2$ be the infinitesimal operators corresponding to the respective parameters, i.e., say the infinitesimals of the group leaving invariant (2.10-3') are of the form

$$\begin{aligned} \xi(x,t) &= a\xi_1(x,t) + b\xi_2(x,t) \\ \tau(x,t) &= a\tau_1(x,t) + b\tau_2(x,t) \\ \pi(x,t,p) &= ag_1(x,t)p + bg_2(x,t)p \end{aligned} \tag{2.10-15}$$

then

$$\mathfrak{X}_1 = \xi_1(x,t)\frac{\partial}{\partial x} + \tau_1(x,t)\frac{\partial}{\partial t} + g_1(x,t)p\frac{\partial}{\partial p} \tag{2.10-16}$$

and

$$\mathfrak{X}_2 = \xi_2(x,t)\,\frac{\partial}{\partial x} + \tau_2(x,t)\,\frac{\partial}{\partial t} + g_2(x,t)p\,\frac{\partial}{\partial p}\,. \tag{2.10-17}$$

If the invariants corresponding to $\mathfrak{X}_1$ and $\mathfrak{X}_2$ are <u>mutually distinct</u>, i.e., functionally independent, then we are able to move directly to the solution of our problem without passing through the intermediate stage of substituting a similarity form into the original partial differential equation and solving the resulting ordinary differential equation. The invariants corresponding to $\mathfrak{X}_1$ and $\mathfrak{X}_2$ are mutually distinct if and only if

$$\mathfrak{X}_1 \neq \lambda(x,t)\mathfrak{X}_2 \tag{2.10-18}$$

for any function $\lambda(x,t)$.

For example,

$$\mathfrak{X}_1 = \frac{\partial}{\partial x} \quad \text{and} \quad \mathfrak{X}_2 = \frac{\partial}{\partial t}$$

have mutually distinct invariants whereas

$$\mathfrak{X}_1 = \frac{\partial}{\partial x} + \frac{\partial}{\partial t} \quad \text{and} \quad \mathfrak{X}_2 = x\,\frac{\partial}{\partial x} + x\,\frac{\partial}{\partial t}$$

have the same invariants even though they correspond to distinctly different group transformations.

Say $\mathfrak{X}_1$ and $\mathfrak{X}_2$ have mutually distinct invariants. Let ζ_i and $F_i(\zeta_i)G_i(x,t)$, $i = 1,2$, be the respective similarity variables and similarity forms. Then if our solution is unique

$$p = P(x,t) = F_1(\zeta_1)G_1(x,t) = F_2(\zeta_2)G_2(x,t). \tag{2.10-19}$$

If ζ_1 is functionally independent of ζ_2, then we express x and t in terms of ζ_1 and ζ_2 in (2.10-19) and solve the resulting functional equation. As a result we would find that

$$P(x,t) = AH(\zeta_1)G_1(x,t) \tag{2.10-20}$$

where the dependence of H on ζ_1 is known explicitly and A is a constant to be determined from the source condition in (2.10-3').

For the infinite bar problem of §2.8, we used

$$\mathfrak{X}_1 = t\frac{\partial}{\partial x} - \frac{x}{2}u\frac{\partial}{\partial u}$$

and

$$\mathfrak{X}_2 = x\frac{\partial}{\partial x} + 2t\frac{\partial}{\partial t} - u\frac{\partial}{\partial u} .$$

Clearly, $\mathfrak{X}_1$ and $\mathfrak{X}_2$ are mutually distinct.

$$\zeta_1 = t, \quad G_1(x,t) = e^{-\frac{x^2}{4t}}$$

$$\zeta_2 = \frac{x}{\sqrt{t}}, \quad G_2(x,t) = \frac{1}{\sqrt{t}}$$

$$F_1(\zeta_1)e^{-\frac{x^2}{4t}} = F_2(\zeta_2)\frac{1}{\sqrt{t}}$$

=> $$F_1(\zeta_1)e^{-\frac{\xi_2^2}{4}} = \frac{F_2(\zeta_2)}{\sqrt{\zeta_1}} . \tag{2.10-21}$$

This functional equation was easily solved. We simply put all the dependence on ζ_1 and ζ_2 respectively on the left and right sides of (2.10-21). Hence

$$\sqrt{\zeta_1}\, F_1(\zeta_1) = F_2(\zeta_2)e^{\frac{\zeta_2^2}{4}} . \tag{2.10-22}$$

Thus

$$F_1(\zeta_1) = \frac{c}{\sqrt{\zeta_1}}$$

where the source condition in the problem was used to show that $c = \dfrac{1}{2\sqrt{\pi}}$.

A general procedure for solving the functional equation in

(2.10-19) is to treat ζ_1 and ζ_2 as independent variables and differentiate each side of the equality with respect to ζ_1, say. As a result a simple linear homogeneous first order ordinary differential equation is obtained for F_1 provided there are no calculation errors and the original problem has a unique solution. In particular this first order differential equation is:

$$\frac{dF_1}{d\zeta_1} = F_1 \frac{\partial}{\partial \zeta_1} (\log \frac{G_2}{G_1}). \tag{2.10-23}$$

We now turn our attention to the problem of finding the Lie group (2.10-8) leaving invariant (2.10-3') and the corresponding forcing functions $f(x)$.

<u>Problem 2.10-1</u>:

Show that

$$\begin{aligned} \tau(x,t) &= \tau(t) \\ \xi(x,t) &= x\,\frac{\tau'(t)}{2} + A(t) \\ g(x,t) &= B(t) - \frac{xf(x)\tau'(t)}{4} - \frac{f(x)A(t)}{2} - \frac{xA'(t)}{2} - \frac{x^2\tau''(t)}{8} \end{aligned} \tag{2.10-24}$$

where $A(t)$, $B(t)$, $\tau(t)$ and $f(x)$ satisfy the equation

$$N_1(x,t) + N_2(x,t) = 0 \tag{2.10-25}$$

with

$$\begin{aligned} N_1(x,t) = {} & \tau'(t)\left[\frac{f^2(x)}{4} + \frac{xf(x)f'(x)}{4} - \frac{f'(x)}{2} - \frac{xf''(x)}{4}\right] \\ & + \frac{\tau''(t)}{4} + \tau'''(t)\left[-\frac{x^2}{8}\right] + B'(t) \end{aligned} \tag{2.10-26}$$

$$N_2(x,t) = A(t)\left[\frac{f(x)f'(x)}{2} - \frac{f''(x)}{2}\right] + A''(t)\left[-\frac{x}{2}\right]. \tag{2.10-27}$$

Imposing the restriction that $f(x)$ be an odd function, we see that

$$N_1(x,t) = N_2(x,t) = 0. \tag{2.10-28}$$

If $\tau(t) \not\equiv 0$, $f(x)$ must satisfy the differential equation:

$$[f^2(x) + xf(x)f'(x) - 2f'(x) - xf''(x)]''' = 0. \tag{2.10-29}$$

Solving (2.10-28,29), we find that $f(x)$ satisfies the Riccati equation

$$2f' - f^2(x) + \beta^2x^2 - \gamma + \frac{16\nu^2 - 1}{x^2} = 0 \tag{2.10-30}$$

and that $\tau(t)$, $B(t)$ satisfy the differential equations

$$\tau'''(t) = 4\beta^2\tau'(t), \quad B'(t) = \frac{\gamma\tau'}{4} - \frac{\tau''}{4} \tag{2.10-31}$$

where β, γ, and ν are arbitrary parameters.

If $A(t) \not\equiv 0$, from (2.10-27) we see that $f(x)$ must satisfy the differential equation:

$$[f(x)f'(x) - f''(x)]'' = 0. \tag{2.10-32}$$

Continuing as in the case of $\tau(t) \not\equiv 0$, we find that $f(x)$ satisfies the Riccati equation

$$2f' - f^2(x) + \beta^2x^2 - \gamma = 0 \tag{2.10-33}$$

and $A(t)$ satisfies

$$A'' = \beta^2A(t). \tag{2.10-34}$$

Thus we see that if $f(x)$ satisfies (2.10-33), which corresponds to (2.10-30) when $\nu^2 = \frac{1}{16}$, then we have the possibility of a larger group leaving invariant (2.10-3').

Invariance of the initial source further restricts $\{\tau,\xi,g\}$, namely,

$$\tau(0) = 0, \quad \xi(x_o,0) = 0, \quad g(x_o,0) = -\frac{\partial\xi}{\partial x}(x_o,0). \tag{2.10-35}$$

<u>Case I</u> $\nu^2 \neq \frac{1}{16}$

If $\nu^2 \neq \frac{1}{16}$, then $A(t) \equiv 0$ and (2.10-35) imposes the following initial conditions on $B(t)$ and $\tau(t)$:

$$\tau(0) = 0, \quad \tau'(0) = 0, \quad B(0) = \frac{x_o^2}{8}\tau''(0). \tag{2.10-36}$$

Solving (2.10-31) subject to the initial conditions (2.10-36) and then substituting this result into (2.10-24) we find that the following one-parameter group leaves invariant (2.10-3'):

$$\begin{aligned} \tau &= \tau_1(t) = 4\sinh^2\beta t \\ \xi &= \xi_1(x,t) = 2\beta x \sinh 2\beta t \\ g &= g_1(x,t) = \gamma\sinh^2\beta t - \beta\sinh 2\beta t(1+xf(x)) \\ &\qquad + \beta^2(x_o^2-x^2\cosh 2\beta t). \end{aligned} \tag{2.10-37}$$

Solving the characteristic equations corresponding to (2.10-37) we find that the similarity variable is

$$\zeta_1(x,t) = \frac{x}{\sqrt{\tau_1(t)}}. \tag{2.10-38}$$

Setting $f(x) = -\frac{2V'(x)}{V(x)}$, we obtain the following similarity form for the solution:

$$p(x,t|x_o) = F_1(\zeta_1)G_1(x,t) \tag{2.10-39}$$

where

$$G_1(x,t) = \frac{V(x)}{\sqrt{\sinh\beta t}}\exp\left[\frac{\gamma t}{4} + \frac{\frac{1}{2}\beta x_o^2}{1-e^{2\beta t}} - \frac{\beta x^2\coth\beta t}{4}\right]. \tag{2.10-40}$$

Case II $\nu^2 = \frac{1}{16}$

In this case (2.10-35) imposes the following initial conditions on $A(t)$, $B(t)$ and $\tau(t)$:

$$\begin{aligned} \tau(0) &= 0 \\ A(0) &= -\frac{x_0}{2}\tau'(0) \\ B(0) &= \frac{x_0}{2}A'(0) + \frac{x_0^2}{8}\tau''(0) - \frac{\tau'(0)}{2} . \end{aligned} \tag{2.10-41}$$

Essentially treating $\tau''(0)$, $A'(0)$ and $A(0)$ as arbitrary constants we find that the following three-parameter (a,b,c) group leaves invariant (2.10-3') when $f(x)$ satisfies (2.10-33):

$$\left.\begin{aligned} \xi(x,t) &= a[2\beta x \sinh 2\beta t] \\ &\quad + b[2 \sinh \beta t] \\ &\quad + c[4\beta(x_0 \cosh \beta t - x \cosh 2\beta t)] \\ \tau(x,t) &= a[4 \sinh^2\beta t] + c[-4 \sinh 2\beta t] \\ g(x,t) &= a[\gamma \sinh^2\beta t - \beta \sinh 2\beta t(1+xf(x)) \\ &\qquad + \beta^2(x_0^2-x^2\cosh 2\beta t)] \\ &\quad + b[\beta(x_0 - x \cosh \beta t) - f(x) \sinh \beta t] \\ &\quad + c[4\beta \cosh^2\beta t + 2\beta f(x)(x \cosh 2\beta t - x_0 \cosh \beta t) \\ &\qquad + 2\beta^2 x(x \sinh 2\beta t - x_0 \sinh \beta t) - \gamma \sinh 2\beta t] \end{aligned}\right\} \tag{2.10-42}$$

Let $\mathfrak{X}_1$, $\mathfrak{X}_2$ and $\mathfrak{X}_3$ be the infinitesimal operators corresponding to the parameters a,b and c respectively. (Note that if $b = c = 0$ then we get the group for $\nu^2 \neq \frac{1}{16}$.) Then it can be shown that

$$\mathfrak{X}_3 = -2 \coth \beta t \ \ \mathfrak{X}_1 + 2\beta(x \operatorname{cosech} \beta t + x_o \coth \beta t)\mathfrak{X}_2. \qquad (2.10\text{-}43)$$

Thus invariants of <u>both</u> $\mathfrak{X}_1$ and $\mathfrak{X}_2$ are also invariants w.r.t. $\mathfrak{X}_3$.

For convenience we find the similarity forms corresponding to $\mathfrak{X}_1$ (parameter a, i.e., b = c = 0) and $\mathfrak{X}_2$ (parameter b, i.e., a = c = 0). The similarity form for $\mathfrak{X}_1$ has already been given (2.10-38,40).

Corresponding to $\mathfrak{X}_2$ we obtain the similarity variable

$$\zeta_2(x,t) = t \qquad (2.10\text{-}44)$$

and similarity form

$$p(x,t|x_o) = F_2(\zeta_2)G_2(x,t) \qquad (2.10\text{-}45)$$

where

$$G_2(x,t) = V(x) \exp\left[\frac{\beta x x_o}{2 \sinh \beta t} - \frac{\beta x^2 \coth \beta t}{4}\right]. \qquad (2.10\text{-}46)$$

Equating the similarity forms (2.10-39,45) we are led to the solution

$$F_2(\zeta_2) = F_2(t) = \frac{D}{\sqrt{\sinh \beta t}} \exp\left[\frac{\gamma t}{4} + \frac{\beta x_o^2}{2(1-e^{2\beta t})}\right]. \qquad (2.10\text{-}47)$$

We now find $V(x)$. The substitution $f(x) = -\frac{2V'(x)}{V(x)}$ transforms (2.10-30) into the following second order linear homogeneous differential equation for $V(x)$:

$$4V'' + \left\{\gamma - \beta^2 x^2 - \frac{(16\nu^2 - 1)}{x^2}\right\}V = 0. \qquad (2.10\text{-}48)$$

The solution of (2.10-48) leading to a realizable probability distribution is

$$V(x) = (\tfrac{1}{2}\,\beta x^2)^{\frac{1}{4}+\nu}\, e^{-\frac{\beta x^2}{4}}\, M(a,b,\tfrac{1}{2}\,\beta x^2) \tag{2.10-49}$$

where $M(a,b,z)$ denotes Kummer's hypergeometric function of the first kind[6] and

$$a = \frac{1}{2} + \nu - \gamma/8\beta, \quad b = 1 + 2\nu \tag{2.10-50}$$

with $\nu > -\frac{1}{2}$, $a \geq 0$. The properties of $M(a,b,z)$ are well known.

As $z \to 0$,

$$M(a,b,z) = 1 + \frac{a}{b}\, z + 0(z^2). \tag{2.10-51}$$

As $z \to +\infty$,

$$M(a,b,z) = \frac{\Gamma(b)}{\Gamma(a)}\, e^z z^{a-b}[1 + 0(z^{-1})]\,. \tag{2.10-52}$$

$$M(0,b,z) \equiv 1. \tag{2.10-53}$$

Using (2.10-51,52) we can show that if $a \neq 0$,

$$\text{(i)} \quad \lim_{x \to \infty} \frac{f(x)}{x} = -\beta\,. \tag{2.10-54}$$

$$\text{(ii)} \quad \lim_{x \to 0} x f(x) = -(4\nu+1). \tag{2.10-55}$$

In order that $\frac{\partial p}{\partial x}$ and p be continuous for $x \in R = (r_1;\, r_2)$, the "directly" generated similarity solutions are of the following types:

(i) $R = (-\infty;\, \infty)$ corresponding to $\nu = -\frac{1}{4}$, (invariance under a three-parameter group).

(ii) $R = (0;\, \infty)$, i.e., $x = 0$ is a reflecting boundary, corresponding to $\nu > -\frac{1}{2}$ (invariance under a one-parameter group).

---

[6] M. Abramowitz and I. A. Stegun, Handbook of Mathematical Functions, Chapter 13, National Bureau of Standards (1964).

Case I $\nu^2 = \frac{1}{16}$

The solution (2.10-45,46,47) has the property that $\frac{\partial p}{\partial x} + f(x)p \neq 0$ for any $x \varepsilon (-\infty; \infty)$. Moreover, $\frac{\partial p}{\partial x}$ is continuous at $x = 0$ iff $\nu = -\frac{1}{4}$. As a result we can directly generate a similarity solution for $R = (-\infty; \infty)$ corresponding to $\nu = -\frac{1}{4}$. By appropriately placing a source at $-x_o$ we are led to solutions for $\nu = \pm\frac{1}{4}$ corresponding to $R = (0; \infty)$. In each case the constant D is determined from the source condition in (2.10-3').

(i) $\nu = -\frac{1}{4}$. In this case

$$V(x) = e^{-\frac{\beta x^2}{4}} M(a,\tfrac{1}{2},\tfrac{1}{2}\beta x^2), \quad a = \frac{1}{4} - \frac{\gamma}{8\beta}. \tag{2.10-56}$$

(a) $R = (-\infty; \infty)$

Here

$$p(x,t|x_o) = p_1(x,t|x_o)$$

$$= \frac{DM(a,\frac{1}{2},\frac{1}{2}\beta x^2)}{\sqrt{\sinh \beta t}} \exp[-\frac{\beta}{4}(1+\coth \beta t)(x-x_o e^{-\beta t})^2] e^{\frac{\gamma t}{4}} \tag{2.10-57}$$

where

$$D = \sqrt{\frac{\beta}{4\pi}} \cdot \frac{1}{M(a,\frac{1}{2},\frac{1}{2}\beta x_o^2)}. \tag{2.10-58}$$

(b) $R = (0; \infty)$

In this case we place sources at $\pm x_o$; the resulting solution is a well-behaved even function of x

$$\Rightarrow \quad \frac{\partial p}{\partial x} + f(x)p = 0 \quad \text{at} \quad x = 0.$$

The solution is

$$p(x,t|x_o) = p_1(x,t|x_o) + p_1(x,t|-x_o). \tag{2.10-59}$$

If $a = 0$, then $f(x) = \beta x$, which corresponds to a free particle in Brownian motion.

(ii) $\nu = \frac{1}{4}$. In this case

$$V(x) = xe^{-\frac{\beta x^2}{4}} M(a,\tfrac{3}{2},\tfrac{1}{2}\beta x^2), \quad a = \frac{3}{4} - \frac{\gamma}{8\beta}. \tag{2.10-60}$$

The corresponding similarity form $p_2(x,t\ x_o)$ leads to a solution obtained by placing sources at $\pm x_o$. The solution corresponds to $R = (0;\ \infty)$

$$p(x,t|x_o) = p_2(x,t|x_o) + p_2(x,t|-x_o) \tag{2.10-61}$$

where

$$p_2(x,t|x_o) =$$

$$= \frac{Dx\ M(a,\frac{3}{2},\frac{1}{2}\beta x^2)}{\sqrt{\sinh \beta t}} \exp[-\frac{\beta}{4}(1+\coth \beta t)(x-x_o e^{-\beta t})^2] e^{\frac{\gamma t}{4}} \tag{2.10-62}$$

with

$$D = \frac{1}{x_o}\sqrt{\frac{\beta}{4\pi}} \cdot \frac{1}{M(a,\frac{3}{2},\frac{1}{2}\beta x_o^2)}. \tag{2.10-63}$$

<u>Case II</u> $\nu > -\frac{1}{2}$

In this case $V(x)$ satisfies (2.10-49), a one-parameter group (2.10-37) leaves invariant (2.10-3') and (2.10-39,40) is the obtained similarity form. Let $\zeta = \zeta_1$ and $F(\zeta) = F_1(\zeta_1)$. Substituting (2.10-39) into the partial differential equation of (2.10-3'), we find that $F(\zeta)$ satisfies a second order linear ordinary differential equation whose general solution can be expressed in terms of Modified Bessel Functions:

$$F(\zeta) = \begin{cases} \zeta^{\frac{1}{2}}[A_1 I_{2\nu}(\kappa\zeta) + A_2 I_{-2\nu}(\kappa\zeta)] & \text{for} \quad x > 0 \\ |\zeta|^{\frac{1}{2}}[B_1 K_{2\nu}(\kappa|\zeta|) + B_2 I_{2\nu}(\kappa|\zeta|)] & \text{for} \quad x < 0. \end{cases} \tag{2.10-64}$$

$\kappa = \beta x_o$ and $\{A_1, A_2, B_1$ and $B_2\}$ are arbitrary constants to be determined by boundary and continuity conditions. As $t \to 0$, $\zeta \to +\infty$. As $z \to +\infty$[7]

$$K_{2\nu}(z) = (\tfrac{\pi}{2z})^{\frac{1}{2}} e^{-z}[1 + O(z^{-1})] \tag{2.10-65}$$

$$I_{2\nu}(z) = (\tfrac{1}{2\pi z})^{\frac{1}{2}} e^{z}[1 + O(z^{-1})]. \tag{2.10-66}$$

In order to have a source only at $x = x_o$, we must set $B_2 = 0$. Continuity of $p(x,t|x_o)$ and the requirement that $\frac{\partial p}{\partial x} + fp$ must never change sign leads us to the conclusion that if $\nu \neq -\frac{1}{4}$, then $R = (0;\ \infty)$. Correspondingly $B_1 = A_2 = 0$ and

$$A_1 = \frac{\beta x_o^{\frac{1}{2}} (\frac{1}{2}\ \beta x_o^2)^{-\frac{1}{4}-\nu}}{M(a,b,\frac{1}{2}\ \beta x_o^2)} . \tag{2.10-67}$$

Next, we show how symmetry aids us in the computation of $\langle x^2 \rangle$. In fact, we will be able to compute $\langle x^2 \rangle$ without recourse to the computation for p.

For <u>all</u> of our computed solutions, (2.10-3') is invariant under (2.10-37). Thus

$$1 = \int_{r_1}^{r_2} p(x,t|x_o)\,dx \tag{2.10-68}$$

is invariant under (2.10-37).

---

(7) G. N. Watson, A Treatise on the Theory of Bessel Functions, Cambridge University Press (1922), 7.23.

Hence $$0 = \int_{r_1}^{r_2} [g(x,t) + \frac{\partial\xi}{\partial x}]p\,dx. \tag{2.10-69}$$

Substituting (2.10-37) into (2.10-69) we find that

$$\beta \sinh 2\beta t \int_{r_1}^{r_2} [xf(x) - 1]p\,dx$$

$$= -\beta^2 \cosh 2\beta t \langle x^2\rangle + \beta^2 x_o^2 + \gamma \sinh^2 \beta t \tag{2.10-70}$$

$$\int_{r_1}^{r_2} [xf(x) - 1]p\,dx = \int_{r_1}^{r_2} x[f(x)p + \frac{\partial p}{\partial x}]dx = -\frac{1}{2}\frac{d}{dt}\langle x^2\rangle \tag{2.10-71}$$

after integration by parts a second time and using the given partial differential equation.

Furthermore, from (2.10-71) we see that

$$(\frac{d}{dt}\langle x^2\rangle)_{t=0} = 2[1 - x_o f(x_o)]. \tag{2.10-72}$$

Substituting (2.10-71) into (2.10-70) we see that $\langle x^2\rangle$ satisfies the first order linear differential equation:

$$\beta \sinh 2\beta t \frac{d}{dt}\langle x^2\rangle$$

$$= 2\beta^2 \cosh 2\beta t \langle x^2\rangle - 2\beta^2 x_o^2 - 2\gamma \sinh^2 \beta t. \tag{2.10-73}$$

The solution of (2.10-72,73) is:

$$\langle x^2\rangle = \frac{4a}{\beta}\frac{M(a+1,b,\frac{1}{2}\beta x_o^2)}{M(a,b,\frac{1}{2}\beta x_o^2)} \sinh 2\beta t + \frac{1}{2}\gamma\frac{(1-e^{-2\beta t})}{\beta^2} + x_o^2 e^{-2\beta t}. \tag{2.10-74}$$

We see that $\langle x^2\rangle$ is bounded iff $a = 0$, i.e.,

$$\gamma = 4\beta(1+2\nu) \tag{2.10-75}$$

and $\beta > 0$.

This corresponds to forcing functions of the form

$$f(x) = \frac{\alpha}{x} + \beta x, \quad \alpha < 1, \quad \beta > 0. \tag{2.10-76}$$

The solution is $(R = (0;\ \infty))$:

$$p(x,t|x_o) = \left[\beta x_o^{\frac{1}{2}} \left(\frac{x}{x_o}\right)^{-\frac{1}{2}\alpha} \zeta^{\frac{1}{2}} I_{-(\frac{1}{2}+\frac{1}{2}\alpha)}(\kappa\zeta) \frac{e^{\frac{(1-\alpha)\beta}{2}t}}{\sqrt{2}\sinh \beta t}\right]$$

$$\cdot \exp\left[-\frac{\beta}{4}(1+\coth \beta t)(x-x_o e^{-\beta t})^2\right] \tag{2.10-77}$$

$$\zeta = \frac{x}{2 \sinh \beta t} .$$

$$\langle x^2 \rangle = \left(\frac{1-\alpha}{\beta}\right)(1-e^{-2\beta t}) + x_o^2 e^{-2\beta t} \tag{2.10-78}$$

and

$$\lim_{t \to \infty} \langle x^2 \rangle = \frac{(1-\alpha)}{\beta} . \tag{2.10-79}$$

For this solution

$0 < \alpha < 1 \leftrightarrow$ "absorbing barrier" at $x = 0$,

$\alpha < 0 \leftrightarrow$ "reflecting barrier" at $x = 0$.

The derived new integrals involving special functions result from

$$\int_R p(x,t|x_o)\,dx = 1.$$

2.11. The Green's Function for an Instantaneous Line Particle Source Diffusing in a Gravitational Field and Under the Influence of a Linear Shear Wind - An Example of a P. D. E. in Three Variables Invariant Under a Two Parameter Group

Neuringer [(1)] found an analytical expression for the distribution function corresponding to an instantaneous line particle source impressed in a gravitational field with a background fluid medium whose velocity field is linearly sheared. The governing equation is the convective form of Smoluchowski's diffusion equation:

$$\frac{\partial p}{\partial t} + \nabla\cdot(\vec{q}p) = D\nabla^2 p - \nabla\cdot(\frac{\vec{F}}{\beta}\, p) \tag{2.11-1}$$

where $p(\vec{r},t)d\vec{r}$ is the probability of finding a particle in the region $[\vec{r};\ \vec{r}+d\vec{r}]$ at time t, D is the diffusion coefficient (assumed constant), $\vec{F}$ is the external force acting on the dispersed particles, $\vec{q}$ is the convective velocity and β is the Stokes viscous drag parameter. For a gravitational field, the external force is $\vec{F} = -c\beta\vec{j}$ where the constant parameter c takes into account buoyancy effects. $\vec{q} = ay\vec{i}$ is the velocity field corresponding to the linear shear wind.

Due to invariance under translation in x, we can assume that the source is located at $x = 0$, $y = y_0$. Note that $\{\frac{c}{a}\} = L$, $\{\frac{c^2}{a^2 D}\} = \{\frac{1}{a}\} = T$. Let $d = \frac{aD}{c^2}$. We non-dimensionalize the problem by making the following change of variables:

$$y' = \frac{a}{c}\, y,\ (y_0' = \frac{a}{c}\, y_0),\ x' = \frac{a}{c}\, x,\ p' = \frac{c^2}{a^2}\, p,\ t' = at. \tag{2.11-2}$$

Dropping the primes, we obtain:

---

(1) Joseph L. Neuringer, "Green's function for an instantaneous line particle source diffusing in a gravitational field and under the influence of a linear shear wind", SIAM J. Appl. Math., 16(1968), pp. 834-842.

$$\frac{\partial p}{\partial t} + y\frac{\partial p}{\partial x} - \frac{\partial p}{\partial y} - d\left(\frac{\partial^2 p}{\partial x^2} + \frac{\partial^2 p}{\partial y^2}\right) = 0 \tag{2.11-3}$$

with the initial condition

$$p(x,y,0) = \delta(x)\delta(y-y_o). \tag{2.11-4}$$

Neuringer found the solution of (2.11-3,4) by first making a change of variables so that the resulting coefficients are independent of the "new" x and y. Using a double Fourier transform in the "new" x and y, he reduced the problem to solving a simple first order ordinary differential equation in the transformed variable. An analytical expression was found for the double inverse transform.

Here we will show that (2.11-3,4) is invariant under a two-parameter Lie group. Using each parameter in turn to generate a functional form for the solution, we are able to reduce (2.11-3) to a first order linear homogeneous ordinary differential equation in the time (t) variable.(2)

Say

$$\begin{aligned}
p^* &= p + \varepsilon g(x,y,t)p + O(\varepsilon^2)\\
t^* &= t + \varepsilon\tau(x,y,t) + O(\varepsilon^2)\\
x^* &= x + \varepsilon\xi(x,y,t) + O(\varepsilon^2)\\
y^* &= y + \varepsilon\eta(x,y,t) + O(\varepsilon^2)
\end{aligned} \tag{2.11-5}$$

leaves invariant (2.11-3).

We transform the various terms of (2.11-3) under (2.11-5) and substitute $-p_{xx} + \frac{1}{d}[p_t+yp_x-p_y]$ for p_{yy} in the resulting $O(\varepsilon)$

(2) Another way of finding the solution of (2.11-3,4): This system represents a stationary, Gaussian, Markov process and moreover corresponds to a linear (Gaussian) Langevin equation. Hence the solution can be expressed in terms of its first and second moments which are easily computable. In particular, these moments are easily calculated from knowledge of the two-parameter group leaving invariant (2.11-3,4) without finding the solution.

expression. Then successively equating to zero the coefficients of p_{xt}, p_{yt}, p_{xy}, p_{xx}, p_t, p_y, p_x and p, we obtain the following equations for the infinitesimals g, τ, ξ, and η:

p_{xt}: $$\frac{\partial \tau}{\partial x} = 0 \qquad (2.11\text{-}6)$$

p_{yt}: $$\frac{\partial \tau}{\partial y} = 0 \qquad (2.11\text{-}7)$$

(2.11-6,7) =>

$$\tau = \tau(t). \qquad (2.11\text{-}8)$$

p_{xy}: $$\frac{\partial \eta}{\partial x} + \frac{\partial \xi}{\partial y} = 0 \qquad (2.11\text{-}9)$$

p_{xx}: $$\frac{\partial \xi}{\partial x} - \frac{\partial \eta}{\partial y} = 0 \qquad (2.11\text{-}10)$$

p_t : $$-\tau'(t) + 2\frac{\partial \eta}{\partial y} = 0 \qquad (2.11\text{-}11)$$

$$(2.11\text{-}11) \Rightarrow \frac{\partial^2 \eta}{\partial y^2} = \frac{\partial^2 \eta}{\partial y \partial x} = 0$$

$$(2.11\text{-}9,10) \Rightarrow \frac{\partial^2 \eta}{\partial x^2} + \frac{\partial^2 \eta}{\partial y^2} = 0$$

Hence

$$\eta = \frac{y}{2}\tau'(t) - x\gamma(t) + \kappa(t) \qquad (2.11\text{-}12)$$

where $\tau(t), \gamma(t)$ and $\kappa(t)$ are arbitrary functions of t. Substituting (2.11-12) into (2.11-9,10), we find that

$$\xi = \frac{x\tau'(t)}{2} + y\gamma(t) + \lambda(t) \qquad (2.11\text{-}13)$$

where $\lambda(t)$ is arbitrary.

p_y: $$\frac{\partial \eta}{\partial t} + y\frac{\partial \eta}{\partial x} + 2d\frac{\partial g}{\partial y} + \frac{\partial \eta}{\partial y} = 0. \qquad (2.11\text{-}14)$$

Substituting (2.11-12,13) into (2.11-14) we see that

$$2dg(x,y,t) = \frac{y^2}{4}[2\gamma - \tau''] + xy\gamma' - \frac{y}{2}[\tau' + 2\kappa'] + \phi(x,t) \qquad (2.11\text{-}15)$$

where $\phi(x,t)$ is an arbitrary function of x and t.

$$p_x: \qquad -\frac{\partial\xi}{\partial t} + \eta - y\frac{\partial\xi}{\partial x} + \frac{\partial\xi}{\partial y} - 2d\frac{\partial g}{\partial x} + 2y\frac{\partial\eta}{\partial y} = 0. \tag{2.11-16}$$

Substituting (2.11-12,13,15) into (2.11-16), we are led to:

$$\gamma(t) = \frac{\tau(t)}{2} + M, \quad M = \text{const.} \tag{2.11-17}$$

and

$$\phi(x,t) = -\frac{x^2}{4}[\tau''+\tau+2M] + x[\tfrac{\tau}{2}+M+\kappa-\lambda'] + K(t) \tag{2.11-18}$$

where $K(t)$ is arbitrary.

Summarizing our results so far, we have:

$$\tau = \tau(t), \quad \xi = \frac{x\tau'}{2} + \frac{y\tau}{2} + My + \lambda(t), \quad \eta = \frac{y\tau'}{2} - \frac{x}{2}[\tau+2M] + \kappa(t),$$

$$2dg = y^2[-\frac{\tau''}{4} + \frac{\tau}{4} + \frac{M}{2}] + y[-\kappa' - \frac{\tau'}{2}] + xy[\frac{\tau'}{2}] \tag{2.11-19}$$

$$+ x^2[-\frac{\tau''}{4} - \frac{\tau}{4} - \frac{M}{2}] + x[\kappa - \lambda' + \frac{\tau}{2} + M] + K(t).$$

Finally, we equate to zero the coefficient of p:

$$p: \quad y^2[-\frac{\tau'''}{4} + \frac{3}{4}\tau'] + y[-\kappa'' + \kappa - \lambda'] + xy[-\frac{\tau}{2} - M]$$

$$+ x^2[-\frac{\tau'''}{4} - \frac{\tau'}{4}] + x[\kappa' - \lambda'' + \frac{\tau''}{2}]$$

$$+ [K' + \kappa' + \frac{\tau'}{2} + d\tau''] = 0. \tag{2.11-20}$$

Equating to zero each of the bracketed terms in (2.11-20) and solving the resulting ordinary differential equation we find that the following 6-parameter group (A,B,C,D,E,F) leaves invariant (2.11-3):

$$\tau = -2A, \quad \xi = B[\tfrac{1}{6}t^3 - t] + \tfrac{1}{2}Ct^2 + Dt + E,$$

$$\eta = \tfrac{1}{2}Bt^2 + Ct + D, \quad g = \frac{1}{2d}\{B[x - yt - \tfrac{1}{2}t^2] + C[-y - t] - D + F\}. \tag{2.11-21}$$

In order to leave invariant the source condition (2.11-4) we must impose the following conditions on τ,ξ,η,g:

$$\tau(t = 0) = \xi(t = 0) = \eta(t = 0) = 0, \; g(0,y_0,0) = 0. \qquad (2.11\text{-}22)$$

Substituting (2.11-22) into (2.11-21) we find that the following two-parameter subgroup (B,C) leaves invariant (2.11-3,4):

$$\tau = 0, \quad \xi = B[\tfrac{1}{6} t^3 - t] + \tfrac{1}{2} Ct^2,$$
$$\eta = \frac{Bt^2}{2} + Ct, \quad g = \frac{B}{2d} [-yt + x - \frac{t^2}{2}] + \frac{C}{2d} [-y - t + y_0]. \qquad (2.11\text{-}23)$$

Next, we derive the similarity forms corresponding to each of the parameters B,C.

Case I $\quad B = 0, \; C \neq 0$

The corresponding characteristic equations are:

$$\frac{dt}{0} = 2\,\frac{dx}{t^2} = \frac{dy}{t} = \frac{dp}{\frac{p}{2d}[-y-t+y_0]} . \qquad (2.11\text{-}24)$$

The first two equalities of (2.11-24) lead to the invariants (similarity variables) t and

$$\zeta_1^* = x - \tfrac{1}{2}\, yt. \qquad (2.11\text{-}25)$$

Substituting these invariants into the third equality of (2.11-24) we obtain the similarity form

$$p = F_1(\zeta_1^*,t)\,e^{-\frac{1}{4d}\left[2y + \frac{(y-y_0)^2}{t}\right]} \qquad (2.11\text{-}26)$$

where F_1 is an arbitrary function of ζ_1^* and t.

Case II $\quad C = 0, \; B \neq 0$

Here the characteristic equations are:

$$\frac{dt}{0} = \frac{dx}{\frac{1}{6}t^3 - t} = \frac{dy}{\frac{1}{2}t^2} = \frac{(dp)}{\frac{p}{4d}[2x - t^2 - 2yt]} . \qquad (2.11\text{-}27)$$

The corresponding invariants are t and

$$\zeta_2^* = x - \frac{yt}{3} + 2\frac{y}{t} . \qquad (2.11\text{-}28)$$

The generated similarity form is:

$$p = F_2(\zeta_2^*, t) e^{-\frac{1}{2d}\left[\frac{y^2}{t} + y - \frac{3x^2}{t^3 - 6t}\right]} . \qquad (2.11\text{-}29)$$

Solving for x and y in terms of ζ_1^* and ζ_2^* we find that

$$x = \frac{2(6-t^2)}{t^2 + 12}\zeta_1^* + \frac{3t^2}{t^2 + 12}\zeta_2^* , \quad y = -\frac{6t}{t^2 + 12}[\zeta_1^* - \zeta_2^*] . \qquad (2.11\text{-}30)$$

Let

$$\zeta_1 = \frac{1}{t^2 + 12}\zeta_1^* , \quad \zeta_2 = \frac{t}{t^2 + 12}\zeta_2^* . \qquad (2.11\text{-}31)$$

Then

$$y = 6[\zeta_2 - t\zeta_1] , \quad x = 3t\zeta_2 + 2(6 - t^2)\zeta_1 , \qquad (2.11\text{-}32)$$

and, moreover, there exist arbitrary functions G_1 and G_2 such that

$$F_1(\zeta_1^*, t) = G_1(\zeta_1, t) , \quad F_2(\zeta_2^*, t) = G_2(\zeta_2, t) .$$

We now equate the similarity forms (2.11-26) and (2.11-29). Using ζ_1, ζ_2 and t as the new variables, we differentiate each side of the equality with respect to ζ_1. Then we find that G_1 satisfies the partial differential equation:

$$d\frac{\partial G_1}{\partial \zeta_1} = G_1[3y_0 - 6\frac{\zeta_1}{t}(t^2 + 12)] \qquad (2.11\text{-}33)$$

$$\Rightarrow \qquad F_1(\zeta_1^*,t) = G_1 = G(t)e^{-\frac{3\zeta_1^2}{dt}(t^2+12) + \frac{3y_o\zeta_1}{d}} \tag{2.11-34}$$

where $G(t)$ is arbitrary.

Hence

$$p(x,y,t) = G(t)e^{\frac{1}{d}\left[-\frac{y^2}{4t} - \frac{y}{2} + \frac{y_o y}{2t} + 3y_o\zeta_1 - \frac{3\zeta_1^2}{t}(t^2+12)\right]}. \tag{2.11-35}$$

Substituting (2.11-35) into (2.11-3) we find that $G(t)$ satisfies the following linear homogeneous first order ordinary differential equation:

$$\frac{dG}{dt} = \left[-\left(\frac{1}{t} + \frac{t}{t^2+12}\right) + \frac{y_o^2}{4d}\left(\frac{1}{t^2} + \frac{3}{t^2+12} - \frac{72}{(t^2+12)^2}\right)\right]G - \frac{1}{4d}G \tag{2.11-36}$$

$$\Rightarrow \qquad G(t) = \frac{C^*}{t\sqrt{t^2+12}}\left[e^{-\frac{t}{4d} - \frac{3ty_o^2}{4d(t^2+12)} - \frac{y_o^2}{4dt}}\right] \tag{2.11-37}$$

where C^* is an arbitrary constant.

Substituting (2.11-37) into (2.11-35) and combining terms we find that

$$p(x,y,t) = \frac{C^{**}}{t\sqrt{1+\frac{t^2}{12}}}\exp\left[-\frac{1}{d}\left\{\frac{[\frac{1}{2}x-\frac{1}{4}(y+y_o)t]^2}{t(1+t^2/12)} + \frac{(y-y_o+t)^2}{4t}\right\}\right] \tag{2.11-38}$$

where C^{**} is an arbitrary constant.

As $t \to 0$,

$$p \to \frac{C^{**}}{t}e^{-\frac{x^2}{4dt} - \frac{(y-y_o)^2}{4dt}}.$$

Hence

$$C^{**} = \frac{1}{4\pi d}. \tag{2.11-39}$$

Problem 2.11-1.

Use the group (2.11-23) to compute $\langle x\rangle$, $\langle y\rangle$, $\langle xy\rangle$, $\langle x^2\rangle$ and $\langle y^2\rangle$ directly without use of the solution (2.11-38).

2.12. Infinite Parameter Groups - Derivation of the Poisson Kernel

All applications treated so far have been concerned with Lie groups with a finite number of parameters. Laplace's equation in two dimensions is invariant under all conformal mappings, an infinite parameter Lie group. This leads to difficulties in applications to boundary value problems when trying to find a nontrivial subgroup leaving invariant the boundary conditions. As an example we derive the Poisson kernel which is the solution $u(r,\theta)$ of

$$\nabla^2 u = \frac{\partial^2 u}{\partial r^2} + \frac{1}{r}\frac{\partial u}{\partial r} + \frac{1}{r^2}\frac{\partial^2 u}{\partial \theta^2} = 0 \tag{2.12-1}$$

with the boundary condition

$$u(1,\theta) = \delta(\theta). \tag{2.12-2}$$

Say the Lie group

$$\begin{aligned} u^* &= u + \varepsilon g(r,\theta)u + O(\varepsilon^2) \\ r^* &= r + \varepsilon R(r,\theta) + O(\varepsilon^2) \\ \theta^* &= \theta + \varepsilon \Theta(r,\theta) + O(\varepsilon^2) \end{aligned} \tag{2.12-3}$$

leaves invariant (2.12-1). Then the determining equations for the infinitesimals g, R, and Θ are:

$$\frac{2}{r}\frac{\partial g}{\partial r} - \nabla^2\left(\frac{R}{r}\right) = 0 \tag{2.12-4}$$

$$\frac{2}{r^2}\frac{\partial g}{\partial \theta} - \nabla^2 \Theta = 0 \tag{2.12-5}$$

$$\frac{\partial R}{\partial r} - \frac{R}{r} - \frac{\partial \Theta}{\partial \theta} = 0 \tag{2.12-6}$$

$$\frac{\partial \Theta}{\partial r} + \frac{1}{r^2}\frac{\partial R}{\partial \theta} = 0 \qquad (2.12\text{-}7)$$

$$\nabla^2 g = 0 \qquad (2.12\text{-}8)$$

(2.12-6,7) =>

$$\nabla^2 \Theta = 0, \quad \nabla^2 \left(\frac{R}{r}\right) = 0. \qquad (2.12\text{-}9)$$

Substituting (2.12-9) into (2.12-4,5) we see that

$$g = \text{constant} = \lambda, \text{ say}. \qquad (2.12\text{-}10)$$

Let

$$S = \frac{R}{r} \,. \qquad (2.12\text{-}11)$$

Then the only restriction on S and Θ is that they are harmonic conjugates of each other. Hence an <u>infinite-parameter</u> Lie group leaves invariant (2.12-1).

Invariance of (2.12-2) imposes the following restrictions on S and Θ:

$$S(1,\theta) = 0 \qquad (2.12\text{-}12)$$

$$\Theta(1,0) = 0 \qquad (2.12\text{-}13)$$

$$\lambda + \frac{\partial \Theta}{\partial \theta}(1,0) = 0. \qquad (2.12\text{-}14)$$

The general solution of (2.12-9) for S is

$$S = F(re^{i\theta}) + G(re^{-i\theta}) \qquad (2.12\text{-}15)$$

where F and G are arbitrary twice differentiable functions of their respective arguments. We assume that F and G are analytic in an annular region centred about the origin containing the unit circle $r = 1$, so that we try a Laurent series expansion for F and G about the origin.

Let $z = re^{i\theta}$, $\bar{z} = re^{-i\theta}$. Then

$$F(z) = \sum_{n=0}^{\infty} a_n z^n + \sum_{n=1}^{\infty} b_n z^{-n}$$

$$G(\bar{z}) = \sum_{n=0}^{\infty} c_n \bar{z}^n + \sum_{n=1}^{\infty} d_n \bar{z}^{-n}$$

where $\{a_n, b_n, c_n, d_n\}$ are arbitrary constants.[1]

The condition (2.12-12) ⟹

$$\left.\begin{array}{l} \left.\begin{array}{l} c_n = -b_n \\ d_n = -a_n \end{array}\right\} n = 1,2,\ldots \\ c_o = -a_o \end{array}\right\} . \qquad (2.12\text{-}16)$$

Hence

$$S = \sum_{n=1}^{\infty} a_n (z^n - \bar{z}^{-n}) + b_n (z^{-n} - \bar{z}^n) \qquad (2.12\text{-}17)$$

where $\{a_n, b_n\}$ are arbitrary parameters.

After using (2.12-6, 7, 13, 14) to solve for $\boxminus$, we are still left with an infinite number of arbitrary parameters. Thus we consider the subgroup for which only $a_1 = a \neq 0$, $b_1 = b \neq 0$, i.e., we consider

$$S = a(z - \frac{1}{\bar{z}}) + b(\frac{1}{z} - \bar{z}). \qquad (2.12\text{-}18)$$

Let $\alpha = -i(a+b)$, $\beta = a - b$, then

$$R = \alpha \sin \Theta[r^2 - 1] + \beta \cos \Theta[r^2 - 1]. \qquad (2.12\text{-}19)$$

(2.12-6,7,19) ⟹

$$\boxminus = -\alpha \cos \Theta[r + \frac{1}{r}] + \beta \sin \Theta[r + \frac{1}{r}] + \mu \qquad (2.12\text{-}20)$$

---

(1) This infinity of constants forms an infinity of parameters of the Lie group.

where μ is an arbitrary constant.

(12.12-13) $\Longrightarrow$ $\mu = 2\alpha$.

(12.12-14) $\Longrightarrow$ $\lambda = -2\beta$.

Hence the following two-parameter (α,β) subgroup leaves invariant (2.12-1,2):

$$\begin{aligned} g &= -2\beta \\ R &= \alpha \sin\Theta(r^2 - 1) + \beta\cos\Theta(r^2 - 1) \\ \boxminus &= \alpha[2 - \cos\Theta(r + \tfrac{1}{r})] + \beta\sin\Theta(r + \tfrac{1}{r}). \end{aligned} \qquad (2.12\text{-}21)$$

The characteristic equations corresponding to $\alpha(\beta = 0)$ are:

$$\frac{du}{0} = \frac{dr}{\sin\Theta(1 - r^2)} = \frac{d\Theta}{(r + \frac{1}{r})\cos\Theta - 2}. \qquad (2.12\text{-}22)$$

The similarity variable ζ is the integral of the differential equation

$$\frac{dr}{d\Theta} = \frac{(1 - r^2)\sin\Theta}{(r + \frac{1}{r})\cos\Theta - 2} \qquad (2.12\text{-}23)$$

which is linear for the dependent variable $y = \cos\Theta$, and independent variable r, namely,

$$(r^2 - 1)\frac{dy}{dr} = (r + \frac{1}{r})y - 2. \qquad (2.12\text{-}24)$$

Solving (2.12-24), we find that the similarity variable is

$$\zeta = \frac{1 - r^2}{1 - 2r\cos\Theta + r^2} \qquad (2.12\text{-}25)$$

so that the resulting similarity form is

$$u = A(\zeta) \qquad (2.12\text{-}26)$$

where $A(\zeta)$ is an arbitrary function of ζ.

Corresponding to the one-parameter (β) subgroup u must satisfy the first order p.d.e.

$$(r^2 - 1)\cos\Theta \frac{\partial u}{\partial r} + (r + \frac{1}{r}) \sin\Theta \frac{\partial u}{\partial \Theta} = -2u. \qquad (2.12\text{-}27)$$

Substituting the similarity form (2.12-26) into (2.12-27) we find that $A(\zeta)$ satisfies the differential equation

$$\left\{[(r^2 - 1)\cos\Theta]\left[\frac{-2r}{1 - 2r\cos\Theta + r^2} + \frac{(1 - r^2)(2\cos\Theta - 2r)}{(1 - 2r\cos\Theta + r^2)^2}\right]\right.$$

$$\left. + \left[\frac{(r + \frac{1}{r})(\sin\Theta)(1 - r^2)(-2r\sin\Theta)}{(1 - 2r\cos\Theta + r^2)^2}\right]\right\} \frac{dA}{d\zeta} + 2A = 0. \qquad (2.12\text{-}28)$$

Collecting terms, and, expressing the coefficients of (2.12-28) in terms of ζ $\Rightarrow$

$$-\zeta \frac{dA}{d\zeta} + A = 0. \qquad (2.12\text{-}29)$$

Hence

$$A(\zeta) = c\zeta \qquad (2.12\text{-}30)$$

where c is an arbitrary constant.

(2.12-2) leads to

$$c = \frac{1}{2\pi}.$$

Hence the Poisson kernel is

$$u(r,\Theta) = \frac{1}{2\pi} \frac{1 - r^2}{1 - 2r\cos\Theta + r^2}. \qquad (2.12\text{-}31)$$

2.13. Far Field of Transonic Flow

The far field of a slender body of revolution is given by a similarity solution. The special reasoning applied in this non-linear problem is instructive and the basic ideas are sketched here. For more details the interested reader should consult Guderley's book.[(1)]

[(1)] K. G. Guderley, Theorie Schallnahe Stromungen, Springer-Verlag, 1957.

If a slender body of revolution flies at the sonic speed, the main part of the flow field can be calculated from the solution of the inviscid compressible flow equation.

If the body shape is given by

$$r = \delta F(x), \quad 0 < x < 1, \tag{2.13-1}$$

where δ is the body thickness ratio and $F(x)$ is the shape function (cf. Fig. 2.13-1)

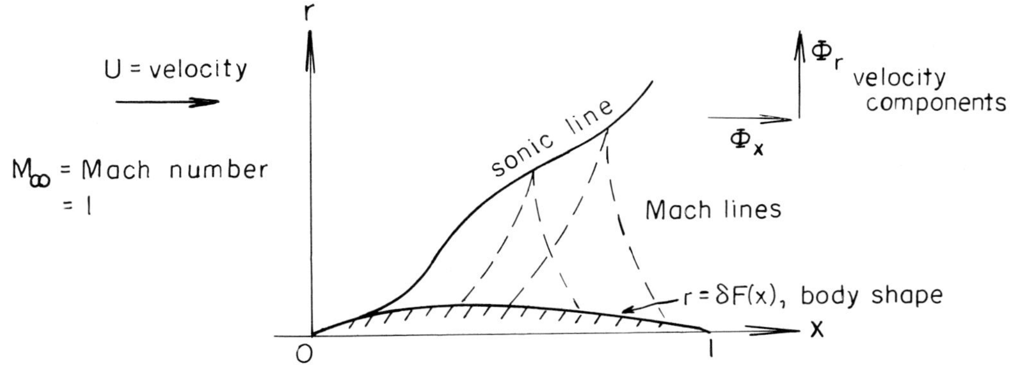

Figure 2.13-1

then the velocity potential can be expanded in an asymptotic perturbation series:

$$\Phi(x,r;\ \delta) = U\{x + \delta^2\phi(x,\tilde{r}) + \cdots\}, \quad \tilde{r} = \delta r. \tag{2.13-2}$$

The contracted coordinated $\tilde{r}$ is used because of the large lateral extent of the perturbation field. The perturbation potential $\phi(x,r)$ satisfies the basic transonic equation

$$(\gamma+1)\phi_x\phi_{xx} = \phi_{\tilde{r}\tilde{r}} + \frac{1}{\tilde{r}}\phi_{\tilde{r}} \tag{2.13-3}$$

(γ = ratio of specific heats = 7/5 for diatomic gases).

The boundary conditions are:

$$\text{tangent flow:} \quad \lim_{\tilde{r} \to 0} \tilde{r}\phi_{\tilde{r}}(x,\tilde{r}) = F(x)F'(x).$$
$$\text{uniform flow at infinity:} \quad \phi \to 0 \quad \text{at} \quad \infty . \tag{2.13-4}$$

(cf. Fig. 2.13-2)

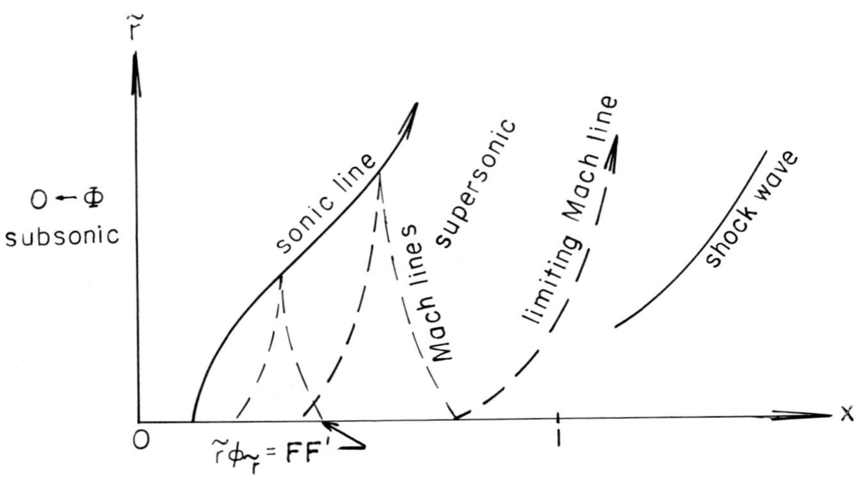

Figure 2.13-2

The equation (2.13-3) is of changing—elliptic in the subsonic region $\phi_x < 0$ and hyperbolic in the supersonic region $\phi_x > 0$. Since the equation is non-linear the location of the sonic line is not known in advance of the solution of any given flow problem. In the hyperbolic region the characteristics, which correspond to Mach waves (weak expansions and compressions) are given by

$$\frac{d\tilde{r}}{dx} = \pm\frac{1}{\sqrt{(\gamma+1)\phi_x}} . \tag{2.13-5}$$

The qualitative features of the solution appear in Fig. 2.13-2. The shock wave respresents a jump in $\phi_x, \phi_{\tilde{r}}$; the jump conditions are found from the integral form of (2.13-3), but this tail shock plays no direct part in the considersations here which concentrate as the flow over the upstream half of the flow field.

The asymptotic field as $\tilde{r} \to \infty$ is of importance both for a basic understanding of transonic flow and for numerical work. In the far field the forward part of the body can be regarded as being concentrated at a point and a solution of similarity form is sought.

The basic equation (2.13-3) is invariant under the stretching transformations:

$$x^* = \alpha x, \quad \tilde{r}^* = \beta\tilde{r}, \quad \phi^* = \alpha^3\beta^{-2}\phi. \tag{2.13-6}$$

This stretching leaves the origin invariant.

There is also translation invariance and x may be replaced by $x - x_o$, since the origin is not really fixed in the far field. By regarding $\beta = \beta(\alpha)$ with $\alpha = \beta = 1$ the identity element we can consider a one parameter group of transformations. From the infinitesimal form of this group we can deduce the required similarity form. If,

$$\alpha = 1 + \varepsilon, \quad \beta(\alpha) = 1 + \beta'(1)\varepsilon = 1 + \frac{1}{k}\,\varepsilon \tag{2.13-7}$$

the infinitesimals of (2.13-6) are

$$\begin{aligned} x^* &= x + \varepsilon\xi; & \xi &= x \\ \tilde{r}^* &= \tilde{r} + \varepsilon\rho; & \rho &= \tilde{r}/k \\ \phi^* &= \phi + \varepsilon\zeta; & \zeta &= (3 - \tfrac{2}{k})\,\phi. \end{aligned} \tag{2.13-8}$$

The characteristic equations of the invariant surface condition are

$$\frac{dx}{x} = k\,\frac{d\tilde{r}}{\tilde{r}} = \frac{d\phi}{(3 - \frac{2}{k})\,\phi}\,. \tag{2.13-9}$$

The integral of the first two of (2.13-9) defines the similarity coordinate

$$\zeta = \frac{x}{\tilde{r}^k} \qquad (2.13\text{-}10)$$

and the integral of the last two of (2.13-9) along the curves ζ = const. yields the desired similarity form

$$\phi(x,\tilde{r}) = \frac{\tilde{r}^{3k-2}}{\gamma+1} f(\zeta) \qquad (2.13\text{-}11)$$

(the factor $\gamma + 1$ is inserted for convenience). The parameter k is unknown at present, but must be found from physical reasoning. The ordinary differential equation for $f(\zeta)$ is

$$(k^2\zeta^2 - \frac{df}{d\zeta}) \frac{d^2f}{d\zeta^2} - k(5k - 4)\zeta \frac{df}{d\zeta} + (3k - 2)^2 f = 0. \qquad (2.13\text{-}12)$$

For $k > 0$, the similarity coordinate covers the $(\tilde{r},x)$ plane with $-\infty < \zeta < \infty$ (cf. Fig. 2.13-3).

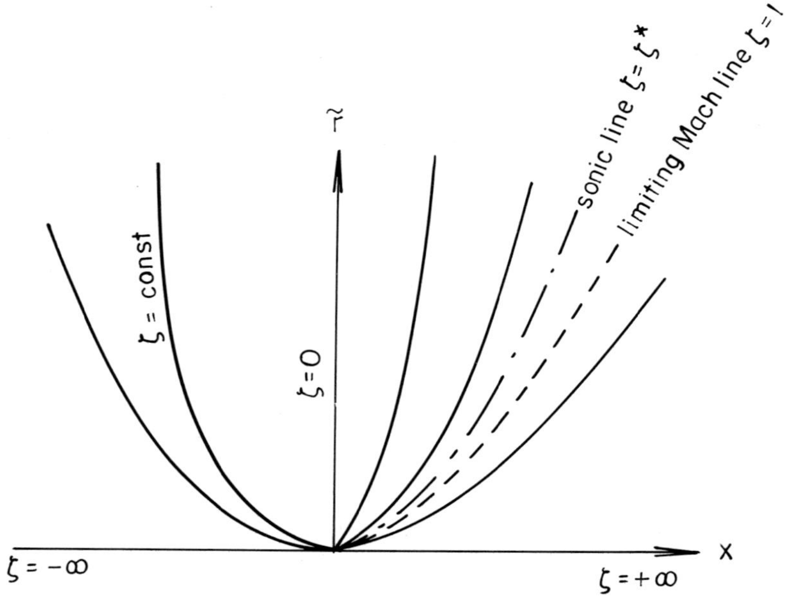

Figure 2-13-3

Note also that

$$\frac{\partial\phi}{\partial x}(x,\tilde{r}) = \frac{\tilde{r}^{2k-2}}{\gamma + 1} f'(\zeta) \qquad (2.13\text{-}13)$$

so that $k < 1$ if we want $\phi_x \to 0$ as $\tilde{r} \to \infty$ along ζ = const. The sonic line has $f'(\zeta) = 0$ at $\zeta = \zeta^*$ for example. Note also that the origin is not really fixed with respect to the near field; in (2.13-12) x can be replaced by $x - x_o$. That the far field should be represented by a singularity near the origin is not a surprising idea, when the far field expansion of linear equations, such as Laplace's equation, is considered. Finally, note that the ordinary differential equation (2.13-12) also admits a group of transformations

$$f^* = \mu^3 f, \quad \zeta^* = \mu\zeta. \qquad (2.13\text{-}14)$$

This has two consequences; first of all by a choice of invariant coordinates as expressed in Part 1, the equation can be reduced to a first-order equation (plus a quadrature); secondly if a solution for any $f(\zeta)$ is found then $\mu^3 f(\frac{\zeta}{\mu})$ is a solution (cf. §2.6). This scale μ cannot be found from far field considerations alone, but has to do with matching to the near field.

There are really no boundary conditions. The solution is sought which is regular as $\zeta \to -\infty$, equivalent to one boundary condition. The essential physically significant condition introduced by Guderley[2] is that there exists a limiting Mach wave, a characteristic which is asymptotic to the sonic line $\zeta = \zeta^*$. The limiting characteristic can be assigned the value $\zeta = 1$ (in view of the remarks above) and that part of the flow $-\infty < \zeta < 1$ is independent of the flow in $\zeta > 1$. The flow in $\zeta > 1$ can be constructed, for example, by the method of characteristics. This condition fixes the value of k and defines the far-field solution. The limiting Mach wave condition becomes, from (2.13-5)

$$\zeta = 1 = \frac{x}{\tilde{r}^k}, \quad \frac{d\tilde{r}}{dx} = \frac{1}{k}\, x^{\frac{1}{k}-1} = \frac{1}{\tilde{r}^{k-1} f'(1)^{1/2}}. \qquad (2.13\text{-}15)$$

---

[2] Ibid.

Thus

$$f'(1) = k^2. \tag{2.13-16}$$

Note that the coefficient of $f''(\zeta)$ in (2.13-13) vanishes at the limiting characteristics. The expected course of $f(\zeta)$ is shown in Fig. 2.13-4.

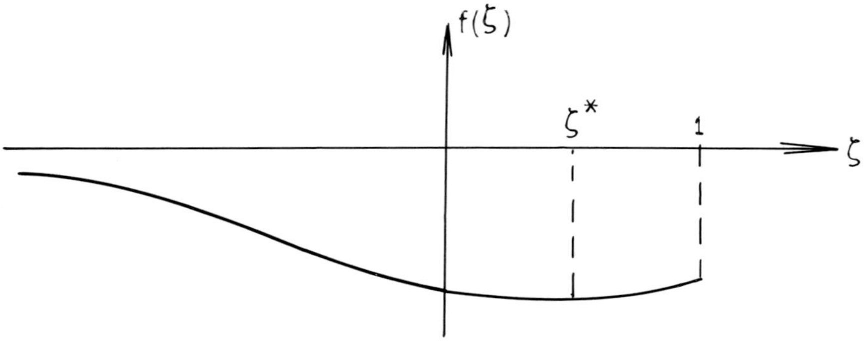

Figure 2.13-4

By studying the expansion of (2.13-13) near $\zeta = -\infty$, we can show that the desired solution has

$$\begin{aligned} f(\zeta) &= a_\infty(-\zeta)^{-(\frac{2}{k}-3)}\{1 + O(\tfrac{1}{\zeta^2})\}, \\ (\gamma+1)\phi(x,\tilde{r}) &\simeq a_\infty \frac{1}{(-x)^{\frac{2}{k}-3}} . \end{aligned} \tag{2.13-17}$$

Physically realistic possibilities must be in the range $0 < k < \frac{2}{3}$. Invariant coordinates corresponding to the group (2.13-14) can be chosen to be

$$s = \frac{1}{\zeta^3} f, \; t = \frac{1}{\zeta^2}\frac{df}{d\zeta} . \tag{2.13-18}$$

Thus

$$\frac{ds}{t-3s} = \frac{d\zeta}{\zeta} \tag{2.13-19}$$

which provides the mapping from an integral curve in the (s,t) plane to the physical coordinate ζ. The first-order differential equation which results from (2.13-12) is

$$\frac{dt}{ds} = \frac{2t^2 - k(4-3k)t - (2-3k)^2 s}{(t-3s)(k^2-t)} . \tag{2.13-20}$$

The singular point at the origin in the (s,t) plane corresponds to $\zeta \to -\infty$ in (2.13-17). The path must be an exceptional one at the origin which then crosses $\zeta \to 0-$ as $s \to +\infty$, $t \to -\infty$ and then reappears at $s \to -\infty$, $t \to -\infty$ as $\zeta \to 0+$. The path must then cross the sonic line at $t = 0$, $s < 0$ and run into the singular point

$$t = k^2, \quad s = -k^3 \frac{4 - 5k}{(2-3k)^2}$$

which corresponds to the limiting Mach wave. (See Fig. 2.13-5). The singularity at the origin is a node while that at the limiting Mach wave is a saddle. All conditions can be met only for one value of k and Guderley showed by numerical integration that $k = 4/7$.

If an integral curve crosses $t = k^2$ at other than the singular point, the mapping to ζ turns around (limit line) and the solution is physically unrealistic. However, later work by Randall[3], Szaniawski[4], Müller and Matschat[5] was able to provide a closed form solution for this exceptional path and show that k is exactly 4/7. The choice $\zeta = 1$ at the singular point of the limiting Mach wave fixes the ζ for the solution by the quadrature (2.13-19).

---

(3) Randall, D. G., Private Communication, 1965.

(4) Szaniawski, A., Two Parametrical Forms of the Self-Similar Transonic Guderley-Frankl Solution, Z.A.M.M., 47:342, No. 5, 1967.

(5) Müller, E. A., and K. Matschat, Ähnlichkeit Lösungen der Transonicher Gleichungen bei der Anstrom Machzahl I.

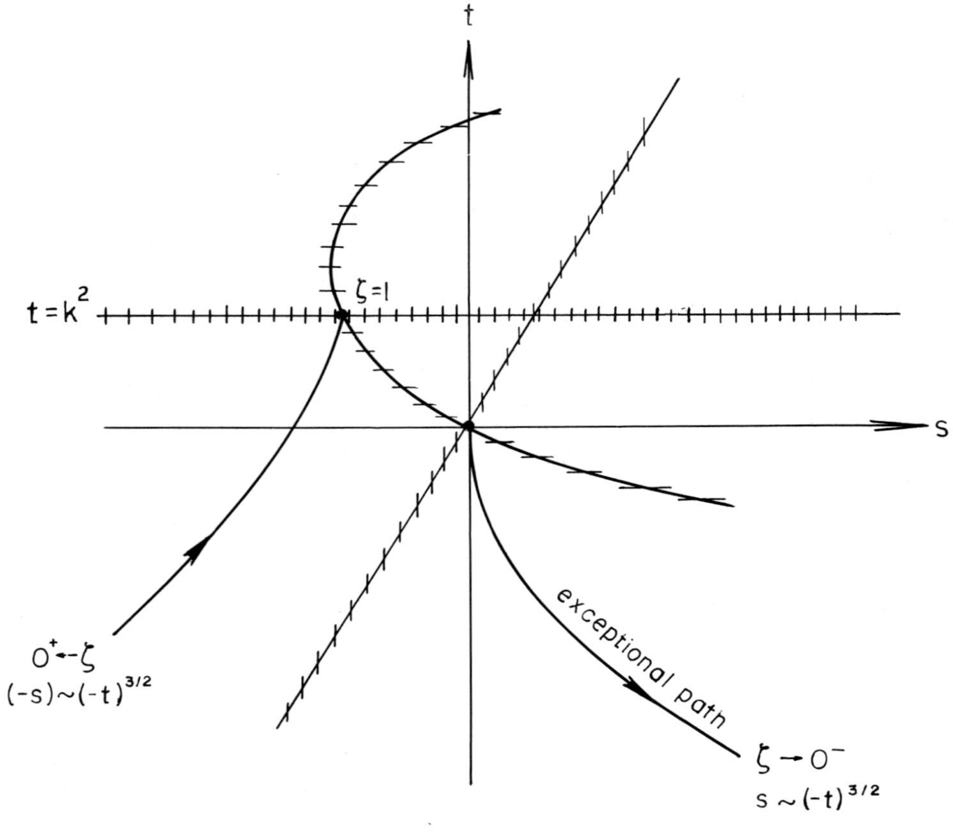

Figure 2.13-5
Phase Plane

As a result of these references Guderley[(6)] and later workers are able to re-interpret this far-field as the first term of the asymptotic field for the exact equation for ϕ, and to develop subsequent terms.

---

(6) Guderley, K. G., and M. C. Breiter, The Development At Infinity of Axi-symmetric Flow Patterns with a Free Stream Mach Number of One, U. S. Air Force Aerospace Research Lab., Dept. 66-0066, April, 1966.

Problem 2.13-1

For the two-dimensional transonic equation of the form (in suitable units)

$$\phi_x \phi_{xx} = \phi_{yy}$$

show that if

$$x^* = x + \varepsilon\xi$$
$$y^* = y + \varepsilon\eta$$
$$u^* = u + \varepsilon\nu$$

then

$$\xi = \alpha x + \beta \qquad \alpha,\beta,\sigma,\mu,\delta,\lambda \text{ arbitrary}$$
$$\eta = \sigma y + \mu$$
$$\nu = (3\alpha-2\sigma)u + \delta y + \lambda$$

Thus carry out the necessary integrations and determine the general similarity form of $\phi(x,y)$.

2.14. Nonlinear and Other Examples

In this section we look at the examples of nonlinear heat conduction and the nonlinear wave equation. The aim is to classify these equations with respect to group invariance.

Nonlinear heat conduction

We consider the equation of nonlinear heat conduction[(1),(2)],

$$\frac{\partial}{\partial x}\left(K(u)\,\frac{\partial u}{\partial x}\right) = \frac{\partial u}{\partial t} \qquad (2.14\text{-}1)$$

---

(1) L. V. Ovsjannikov, Gruppovye Svoystva Uravnenya Nelinaynoy Teploprovodnosty, Dok. Akad. Nauk. CCCP, 1959, 125, 3, p. 492.

(2) G. W. Bluman, Construction of Solutions to Partial Differential Equations by the Use of Transformation Groups, Ph.D. Thesis, California Institute of Technology, 1967.

with boundary conditions

$$u(x,0) = 0, \tag{2.14-2}$$

$$u(0,t) = f(t). \tag{2.14-3}$$

$K(u)$ represents a nonlinear heat conductivity, i.e., $K(u) \neq$ const. We find the groups leaving invariant (2.14-1) and the corresponding $\{K(u)\}$ and in turn find the subgroups and $\{f(t)\}$ for which the subgroups leave invariant (2.14-2,3).

First we consider the invariance properties of (2.14-1). Say the Lie group of transformations

$$\begin{aligned} u^* &= u + \varepsilon\eta(x,t,u) + O(\varepsilon^2) \\ t^* &= t + \varepsilon\tau(x,t,u) + O(\varepsilon^2) \\ x^* &= x + \varepsilon\xi(x,t,u) + O(\varepsilon^2) \end{aligned} \tag{2.14-4}$$

leaves invariant (2.14-1). Then it is easy to show that for any $K(u)$

$$\tau_u = \xi_u = \tau_x = 0. \tag{2.14-5}$$

The remaining determining equations are

$$\xi_t + 2K'(u)\eta_x - K(u)\xi_{xx} + 2K(u)\eta_{xu} = 0 \tag{2.14-6}$$

$$K(u)\tau'(t) + \eta K'(u) - 2K(u)\xi_x = 0 \tag{2.14-7}$$

$$K'(u)\tau'(t) + K'(u)\eta_u + \eta K''(u) - 2K'(u)\xi_x + K(u)\eta_{uu} = 0 \tag{2.14-8}$$

$$K(u)\eta_{xx} - \eta_t = 0. \tag{2.14-9}$$

Solving (2.14-7) for η and substituting this result into (2.14-9) we find that

$$\xi = \frac{x}{2}\tau'(t) + ax^2 + bx + c, \quad \tau = \tau(t), \quad \eta = \frac{K(u)}{K'(u)}(4ax+2b) \tag{2.14-10}$$

where a,b,c are arbitrary constants. Substituting (2.14-10) into (2.14-8) we find that if one of a, $b \neq 0$ then necessarily

$$\left(\frac{K}{K'}\right)'' = 0 \tag{2.14-11}$$

$\Longrightarrow$

$$K(u) = \lambda(u+\kappa)^{\nu} \tag{2.14-12}$$

where λ, κ, ν are arbitrary constants if one of $a, b \neq 0$.

Finally substituting (2.14-10) into (2.14-6) we obtain

$$\frac{x}{2}\tau''(t) + 2aK(u)\left[7 - \frac{4K(u)K''(u)}{(K'(u))^2}\right] = 0. \tag{2.14-13}$$

Three cases arise.

Case I (invariance under a three-parameter (α,β,γ) group)

<u>$K(u)$ arbitrary</u>

$$\xi = \beta x + \gamma, \quad \tau = 2\alpha + 2\beta t, \quad \eta = 0. \tag{2.14-14}$$

Case II (invariance under a four-parameter $(\alpha,\beta,\gamma,\delta)$ group)

<u>$K(u) = \lambda(u+\kappa)^{\nu}$</u>

$$\xi = \beta x + \gamma + \delta x, \quad \tau = 2\alpha + 2\beta t, \quad \eta = \frac{2\delta}{\nu}(u+\kappa). \tag{2.14-15}$$

(A limiting case here is $K(u) = \lambda e^{\nu u}$.)

Case III (invariance under a five-parameter $(\alpha,\beta,\gamma,\delta,\rho)$ group)

<u>$K(u) = \lambda(u+\kappa)^{-\frac{4}{3}}$</u>

$$\begin{aligned} \xi &= \beta x + \gamma + \delta x + \rho x^2 \\ \tau &= 2\alpha + 2\beta t \\ \eta &= -\frac{3}{2}\delta(u+\kappa) - 3\rho x(u+\kappa). \end{aligned} \tag{2.14-16}$$

Now we consider the boundary conditions (2.14-2,3):

Invariance of $t = 0 \Rightarrow \alpha = 0$.

Invariance of $x = 0 \Rightarrow \gamma = 0$.

Invariance of (2.14-2) $\Rightarrow \kappa = 0$ in Cases II and III.

Invariance of (2.14-3) $\Rightarrow$

$$\eta(0,t,f(t)) = \tau f'(t). \tag{2.14-17}$$

Hence

Case I: $f(t) = \text{constant} = 1$, say.

$$\text{Case II:} \quad f(t) = Ct^B \tag{2.14-18}$$

where B,C are arbitrary constants. $(\delta = \beta\nu B)$

$$\text{Case III:} \quad f(t) = Ct^B \tag{2.14-19}$$

where B,C are arbitrary constants $(\delta = \beta\nu B)$. In this case the problem appears to be invariant under a two-parameter subgroup (β,ρ). However, this case is meaningless since the corresponding differential equation does not make sense when $u = 0$.

We now examine Cases I and II in more detail.

Case I $K(u)$ arbitrary, $f(t) = 1$.

The characteristic equations are:

$$\frac{dx}{x} = \frac{dt}{2t} = \frac{du}{0} \tag{2.14-20}$$

$\Rightarrow$ the similarity variable is

$$\zeta = \frac{x}{\sqrt{t}} \tag{2.14-21}$$

$(t = 0 \leftrightarrow \zeta = \infty,\ x = 0 \leftrightarrow \zeta = 0)$

and similarity form for the solution

$$u = F(\zeta). \tag{2.14-22}$$

Substituting (2.14-22) into (2.14-1), we obtain the following ordinary differential equation for $F(\zeta)$:

$$\frac{d}{d\zeta}\left(K(F)\frac{dF}{d\zeta}\right) + \frac{\zeta}{2}\frac{dF}{d\zeta} = 0. \qquad (2.14\text{-}23)$$

The corresponding boundary conditions for $F(\zeta)$ are:

$$F(\infty) = 0, \quad F(0) = 1.$$

<u>Case II</u> $K(u) = \lambda u^{\nu}$, $f(t) = Ct^{B}$.

The characteristic equations are:

$$\frac{dx}{(1+\nu B)x} = \frac{dt}{2t} = \frac{du}{2Bu} \qquad (2.14\text{-}24)$$

=> the similarity variable is

$$\zeta = \frac{x}{t^{\frac{\nu B+1}{2}}}. \qquad (2.14\text{-}25)$$

(Assuming $\nu B > -1$, $t = 0 \leftrightarrow \zeta = \infty$, $x = 0 \leftrightarrow \zeta = 0$.)

The corresponding similarity form for the solution is

$$u = t^{B}F(\zeta). \qquad (2.14\text{-}26)$$

Substituting (2.14-26) into (2.14-1), we find that $F(\zeta)$ satisfies the o.d.e.:

$$\lambda\frac{d}{d\zeta}\left(F^{\nu}\frac{dF}{d\zeta}\right) + \frac{\nu B+1}{2}\zeta\frac{dF}{d\zeta} - BF = 0. \qquad (2.14\text{-}27)$$

$F(\zeta)$ satisfies the boundary conditions:

$$F(\infty) = 0 \quad (B > 0), \quad F(0) = C. \qquad (2.14\text{-}28)$$

Note that the one-parameter (μ) group of stretchings

$$\zeta^{*} = \mu\zeta, \quad F^{*} = \mu^{2/\nu}F \qquad (2.14\text{-}29)$$

leaves invariant (2.14-27).

We show how invariance of (2.14-27) under (2.14-29) can be used to convert the given boundary value problem to an initial value problem.

Let

$$z = \frac{1}{\zeta}, \quad G(z) = F(\frac{1}{z}). \tag{2.14-30}$$

Then (2.14-27,28) become

$$\lambda z^2 \frac{d}{dz}(G^\nu z^2 \frac{dG}{dz}) - (\frac{\nu B+1}{2}) z \frac{dG}{dz} - BG = 0 \tag{2.14-31}$$

with boundary conditions

$$G(0) = 0, \quad G(\infty) = C. \tag{2.14-32}$$

(2.14-31) is invariant under the Lie group

$$z^* = \mu z, \ G^* = \mu^{-2/\nu} G. \tag{2.14-33}$$

Say $G = g(z)$ is a solution of (2.14-31) satisfying the initial conditions

$$g(0) = 0, \ g'(0) = 1. \tag{2.14-34}$$

Integrating this solution out to ∞, say

$$g(\infty) = g_\infty \neq 0. \tag{2.14-35}$$

Invariance of (2.14-31) under (2.14-33)

=> for any μ (cf. §2.6)

$$G(z) = \mu^{2/\nu} g(\mu z) \tag{2.14-36}$$

solves (2.14-31).

$$G(\infty) = \mu^{2/\nu} g_\infty, \ G(0) = 0.$$

Choosing μ to be such that

$$\mu = (\frac{c}{g_\infty})^{\nu/2} \tag{2.14-37}$$

we see that (2.14-36) solves (2.14-31,32).

Nonlinear wave equation

Problem 2.14-1

Consider the nonlinear wave equation

$$\frac{\partial^2 u}{\partial t^2} - c^2(u)\frac{\partial^2 u}{\partial x^2} = 0 \tag{2.14-38}$$

where $c(u) \neq$ const. Say the Lie group

$$\begin{aligned} u^* &= u + \varepsilon\eta(x,t,u) + 0(\varepsilon^2) \\ x^* &= x + \varepsilon\xi(x,t,u) + 0(\varepsilon^2) \\ t^* &= t + \varepsilon\tau(x,t,u) + 0(\varepsilon^2) \end{aligned} \tag{2.14-39}$$

leaves invariant (2.14-38). Show that

(i) $c(u)$ arbitrary $\Longrightarrow$

$$\begin{aligned} \eta &= 0 \\ \xi &= \alpha_1 x + \alpha_2 \\ \tau &= \alpha_1 t + \alpha_3 \end{aligned} \tag{2.14-40}$$

(ii) $c(u) = \lambda(u+\kappa)^\nu \Longrightarrow$

$$\begin{aligned} \eta &= \alpha_4(u+\kappa) \\ \xi &= \alpha_1 x + \alpha_2 + \alpha_4 \nu x \\ \tau &= \alpha_1 t + \alpha_3 \end{aligned} \tag{2.14-41}$$

(iii) $c(u) = \lambda(u+\kappa)^2 \implies$

$$\eta = \alpha_4(u+\kappa) + \alpha_5 x(u+\kappa)$$
$$\xi = \alpha_1 x + \alpha_2 + 2\alpha_4 x + \alpha_5 x^2 \qquad (2.14\text{-}42)$$
$$\tau = \alpha_1 t + \alpha_3$$

where $\alpha_1, \alpha_2, \ldots, \alpha_5$ are arbitrary constants.

Other examples

Problem 2.14-2

Consider Laplace's equation in $n \geq 3$ dimensions:

$$\sum_{n=1}^{n} \frac{\partial^2 u}{\partial x_i^2} = 0. \qquad (2.14\text{-}43)$$

(i) Show that if the Lie group

$$u^* = u + \varepsilon u f(x_1, x_2, \ldots, x_n)$$
$$x_i^* = x_i + \varepsilon \xi_i(x_1, x_2, \ldots, x_n), \quad i = 1,2,\ldots,n \qquad (2.14\text{-}44)$$

leaves invariant (2.14-43) then

$$f = (2-n) \sum_{k=1}^{n} \gamma_k x_k + \delta$$

$$\xi_i = \alpha_i + \sum_{j=1}^{n} \beta_{ij} x_j - \gamma_i \sum_{j=1}^{n} x_j^2 + 2x_i \sum_{j=1}^{n} \gamma_j x_j + \lambda x_i \qquad (2.14\text{-}45)$$

$i = 1,2,\ldots,n$

where $\beta_{ij} = -\beta_{ji}$, $i, j = 1,2,\ldots,n$.

The subgroup corresponding to $\delta = 0$ is called the conformal group and contains $\frac{(n+1)(n+2)}{2}$ parameters.

(ii) Consider the Lie group (2.14-45) in $n = 2$ dimensions. Prove that in the (x_1, x_2) plane this group is the well-known Möbius (bilinear) group, i.e., the six-parameter group of transformations

$$\ell(z) = \frac{az + b}{cz + d}, \qquad ad - bc \neq 0$$

where $z = x_1 + ix_2$.

2.15. Construction of Partial Differential Equations Invariant Under a Given Multi-parameter Group

In this section, we consider the problem of finding the most general partial differential equation invariant under a given multi-parameter Lie group. We restrict ourselves to second order partial differential equations with two independent variables but the method applies in general. The construction of all ordinary differential equations of first or second order invariant under a given one-parameter group was discussed previously in §1.15 and §1.17.

Construction of Partial Differential Equations Invariant Under Subgroups of the Group of the Heat Equation

Consider the infinitesimal operators (cf. (2.7-14)) of the Lie group leaving invariant the heat equation. From the commutator table we see that invariance under subalgebras generated by the following operators can be considered:

$$\{\mathfrak{X}_1\},\ \{\mathfrak{X}_2\},\ \{\mathfrak{X}_3\},\ \{\mathfrak{X}_4\},\ \{\mathfrak{X}_5\},\ \{\mathfrak{X}_6\},\ \{\mathfrak{X}_1, \mathfrak{X}_2\},$$

$$\{\mathfrak{X}_1, \mathfrak{X}_2, \mathfrak{X}_3\},\ \{\mathfrak{X}_1, \mathfrak{X}_2, \mathfrak{X}_4\},\ \{\mathfrak{X}_1, \mathfrak{X}_2, \mathfrak{X}_3, \mathfrak{X}_4\},$$

$$\{\mathfrak{X}_1, \mathfrak{X}_3\},\ \{\mathfrak{X}_1, \mathfrak{X}_3, \mathfrak{X}_4\},\ \{\mathfrak{X}_1, \mathfrak{X}_4\},\ \{\mathfrak{X}_1, \mathfrak{X}_4, \mathfrak{X}_5\},$$

$\{ \mathfrak{X}_1, \mathfrak{X}_4, \mathfrak{X}_6\}$, $\{\mathfrak{X}_1, \mathfrak{X}_5\}$, $\{ \mathfrak{X}_1, \mathfrak{X}_6\}$, $\{ \mathfrak{X}_2, \mathfrak{X}_3\}$,

$\{ \mathfrak{X}_2, \mathfrak{X}_3, \mathfrak{X}_4\}$, $\{\mathfrak{X}_2, \mathfrak{X}_3, \mathfrak{X}_5\}$, $\{ \mathfrak{X}_2, \mathfrak{X}_4\}$, $\{ \mathfrak{X}_2, \mathfrak{X}_4, \mathfrak{X}_6\}$,

$\{ \mathfrak{X}_2, \mathfrak{X}_5\}$, $\{ \mathfrak{X}_2, \mathfrak{X}_6\}$, $\{ \mathfrak{X}_3, \mathfrak{X}_4\}$, $\{ \mathfrak{X}_3, \mathfrak{X}_4, \mathfrak{X}_5\}$,

$\{ \mathfrak{X}_3, \mathfrak{X}_4, \mathfrak{X}_6\}$, $\{ \mathfrak{X}_3, \mathfrak{X}_5\}$, $\{ \mathfrak{X}_3, \mathfrak{X}_5, \mathfrak{X}_6\}$, $\{ \mathfrak{X}_3, \mathfrak{X}_6\}$,

$\{ \mathfrak{X}_4, \mathfrak{X}_5\}$, $\{ \mathfrak{X}_4, \mathfrak{X}_5, \mathfrak{X}_6\}$, $\{ \mathfrak{X}_4, \mathfrak{X}_6\}$, $\{ \mathfrak{X}_5, \mathfrak{X}_6\}$.

Each of these thirty-four subalgebras is a Lie algebra and these are the only subalgebras[(1)] of the Lie algebra of the group leaving invariant the heat equation. Recall that the Lie algebra of the heat equation is generated by $\{ \mathfrak{X}_1, \mathfrak{X}_2, \mathfrak{X}_5\}$. Various chains of subalgebras can be constructed, for example

$$\text{(i)}\quad \{ \mathfrak{X}_1\} \subset \{ \mathfrak{X}_1, \mathfrak{X}_2\} \subset \{ \mathfrak{X}_1, \mathfrak{X}_2, \mathfrak{X}_3\} \subset \{ \mathfrak{X}_2, \mathfrak{X}_6\} \subset \{ \mathfrak{X}_2, \mathfrak{X}_4, \mathfrak{X}_6\} \subset \{ \mathfrak{X}_1, \mathfrak{X}_2, \mathfrak{X}_5\} \tag{2.15-1}$$

$$\text{(ii)}\quad \{ \mathfrak{X}_3\} \subset \{ \mathfrak{X}_1, \mathfrak{X}_3\} \subset \{ \mathfrak{X}_1, \mathfrak{X}_3, \mathfrak{X}_4\} \subset \{ \mathfrak{X}_1, \mathfrak{X}_4, \mathfrak{X}_6\} \subset \{ \mathfrak{X}_1, \mathfrak{X}_4, \mathfrak{X}_5\} \subset \{ \mathfrak{X}_1, \mathfrak{X}_2, \mathfrak{X}_5\}. \tag{2.15-2}$$

Each subalgebra leads to a class of partial differential equations invariant under it. If subalgebras $\mathscr{A}$ and $\mathscr{B}$ are members of the same chain, $\mathscr{A} \subset \mathscr{B}$, then a partial differential equation invariant under subalgebra $\mathscr{B}$ is also invariant under subalgebra $\mathscr{A}$.

We now construct the most general partial differential equation of the form

$$u_{xx} = H(x,t,u,u_x,u_t,u_{tt},u_{xt}) \tag{2.15-3}$$

invariant under a given Lie algebra. We illustrate the method by

---

(1) A <u>subalgebra</u>, $\mathscr{L}^*$, of a given Lie algebra $\mathscr{L}$ is a subset of $\mathscr{L}$ such that $\mathscr{L}^*$ forms a Lie algebra under the operations allowed in $\mathscr{L}$.

considering each member of the chain (2.15-1). Before proceeding we must calculate the infinitesimals of the extended transformations (cf. (2.1-19) et seq.) $\{\eta_x, \eta_t, \eta_{xx}, \eta_{tt}, \eta_{xt}\}$ for the group (2.7-12) of the heat equation,

$$\eta_x = \frac{\partial \eta}{\partial x} + \left(\frac{\partial \eta}{\partial u} - \frac{\partial \xi}{\partial x}\right)u_x$$

$$= -\gamma\{\frac{x}{2} u + [\frac{x^2}{4} + \frac{3}{2} t]u_x\} - \delta[\frac{1}{2} u + \frac{1}{2} xu_x] + \lambda u_x - \beta u_x$$

$$\eta_t = \frac{\partial \eta}{\partial t} + \left(\frac{\partial \eta}{\partial u} - \frac{\partial \tau}{\partial t}\right)u_t - \frac{\partial \xi}{\partial t} u_x$$

$$= -\gamma\{\frac{1}{2} u + [\frac{x^2}{4} + \frac{5}{2} t]u_t + xu_x\} - \delta\{\frac{1}{2} xu_t + u_x\} + \lambda u_t - 2\beta u_t$$

$$\eta_{xx} = \frac{\partial^2 \eta}{\partial x^2} + 2 \frac{\partial^2 \eta}{\partial x \partial u} u_x + \left(\frac{\partial \eta}{\partial u} - 2 \frac{\partial \xi}{\partial x}\right)u_{xx}$$

$$= -\gamma\{\frac{1}{2} u + xu_x + [\frac{1}{4} x^2 + \frac{5}{2} t]u_{xx}\} - \delta\{u_x + \frac{1}{2} xu_{xx}\} + \lambda u_{xx} - 2\beta u_{xx}$$

$$\eta_{tt} = \left(2 \frac{\partial^2 \eta}{\partial t \partial u} - \frac{\partial^2 \tau}{\partial t^2}\right) u_t + \left(\frac{\partial \eta}{\partial u} - 2 \frac{\partial \tau}{\partial t}\right)u_{tt} - 2 \frac{\partial \xi}{\partial t} u_{xt}$$

$$= -\gamma\{3u_t + [\frac{1}{4} x^2 + \frac{9}{2} t]u_{tt} + 2xu_{xt}\} - \delta\{\frac{1}{2} xu_{tt} + 2u_{xt}\} + \lambda u_{tt} - 4\beta u_{tt}$$

$$\eta_{xt} = \frac{\partial^2 \eta}{\partial x \partial u} u_t + \left(\frac{\partial^2 \eta}{\partial t \partial u} - \frac{\partial^2 \xi}{\partial t \partial x}\right)u_x + \left(\frac{\partial \eta}{\partial u} - \frac{\partial \xi}{\partial x} - \frac{\partial \tau}{\partial t}\right)u_{xt} - \frac{\partial \xi}{\partial t} u_{xx}$$

$$= -\gamma\{\frac{1}{2} xu_t + \frac{3}{2} u_x + [\frac{1}{4} x^2 + \frac{7}{2} t]u_{xt} + xu_{xx}\}$$

$$-\delta\{\frac{1}{2} u_t + \frac{1}{2} xu_{xt} + u_{xx}\} + \lambda u_{xt} - 3\beta u_{xt}. \qquad (2.15\text{-}4)$$

From the point of view of finding a partial differential equation invariant under a given group, H in (2.15-3) must be considered as a function of seven variables, namely:

$$\left.\begin{aligned} &x_1 = x,\ x_2 = t,\ x_3 = u,\ x_4 = u_x,\ x_5 = u_t, \\ &x_6 = u_{tt} \text{ and } x_7 = u_{xt}. \\ \text{Let}\quad &x_8 = u_{xx}. \end{aligned}\right\} \qquad (2.15\text{-}5)$$

Let X_i be the infinitesimal transformation associated with x_i, i.e.,

$$x_i^* = x_i + \varepsilon X_i + O(\varepsilon^2), \quad i = 1,2,\ldots,8.$$

Then for the group (2.7-12)

$$\begin{aligned}
X_1 &= \kappa + \gamma x_1 x_2 + \delta x_2 + \beta x_1 \\
X_2 &= \alpha + \gamma x_2^2 + 2\beta x_2 \\
X_3 &= -\gamma[\tfrac{1}{4} x_1^2 + \tfrac{1}{2} x_2]x_3 - \tfrac{1}{2} \delta x_1 x_3 + \lambda x_3 \\
X_4 &= -\gamma\{\tfrac{1}{2} x_1 x_3 + [\tfrac{1}{4} x_1^2 + \tfrac{3}{2} x_2]x_4\} - \tfrac{1}{2} \delta[x_3 + x_1 x_4] + \lambda x_4 - \beta x_4 \\
X_5 &= -\gamma\{\tfrac{1}{2} x_3 + x_1 x_4 + [\tfrac{1}{4} x_1^2 + \tfrac{5}{2} x_2]x_5\} - \delta[x_4 + \tfrac{1}{2} x_1 x_5] \\
&\qquad + \lambda x_5 - 2\beta x_5 \\
X_6 &= -\gamma\{3x_5 + [\tfrac{1}{4} x_1^2 + \tfrac{9}{2} x_2]x_6 + 2x_1 x_7\} - \delta[\tfrac{1}{2} x_1 x_6 + 2x_7] \\
&\qquad + \lambda x_6 - 4\beta x_6 \\
X_7 &= -\gamma\{\tfrac{3}{2} x_4 + \tfrac{1}{2} x_1 x_5 + [\tfrac{1}{4} x_1^2 + \tfrac{7}{2} x_2]x_7 + x_1 x_8\} \\
&\quad -\delta[\tfrac{1}{2} x_5 + \tfrac{1}{2} x_1 x_7 + x_8] + \lambda x_7 - 3\beta x_7 \\
X_8 &= -\gamma\{\tfrac{1}{2} x_3 + x_1 x_4 + [\tfrac{1}{4} x_1^2 + \tfrac{5}{2} x_2]x_8\} \\
&\quad - \delta[x_4 + \tfrac{1}{2} x_1 x_8] + \lambda x_8 - 2\beta x_8.
\end{aligned} \tag{2.15-6}$$

Let $\tilde{\tilde{\mathfrak{X}}}_i$, $i = 1,2,\ldots,6$, denote the twice extended infinitesimal operators corresponding to the parameters $\kappa,\alpha,\lambda,\beta,\gamma$, and δ respectively (cf. (2.7-14)). Then from (2.15-6)

$$\tilde{\tilde{\mathfrak{X}}}_1 = \frac{\partial}{\partial x_1}$$

$$\tilde{\tilde{\mathfrak{X}}}_2 = \frac{\partial}{\partial x_2}$$

$$\tilde{\tilde{\mathfrak{X}}}_3 = \sum_{j=3}^{8} x_j \frac{\partial}{\partial x_j}$$

$$\tilde{\tilde{\mathfrak{X}}}_4 = x_1 \frac{\partial}{\partial x_1} + 2x_2 \frac{\partial}{\partial x_2}$$
$$- [x_4 \frac{\partial}{\partial x_4} + 2x_5 \frac{\partial}{\partial x_5} + 4x_6 \frac{\partial}{\partial x_6} + 3x_7 \frac{\partial}{\partial x_7} + 2x_8 \frac{\partial}{\partial x_8}]$$

$$\tilde{\tilde{\mathfrak{X}}}_5 = x_1x_2 \frac{\partial}{\partial x_1} + x_2^2 \frac{\partial}{\partial x_2} - \{[\frac{1}{4} x_1^2 + \frac{1}{2} x_2]x_3 \frac{\partial}{\partial x_3}$$
$$+ [\frac{1}{2} x_1x_3 + \frac{1}{4} x_1^2x_4 + \frac{3}{2} x_2x_4] \frac{\partial}{\partial x_4}$$
$$+ [\frac{1}{2} x_3 + x_1x_4 + \frac{1}{4} x_1^2x_5 + \frac{5}{2} x_2x_5] \frac{\partial}{\partial x_5}$$
$$+ [3x_5 + \frac{1}{4} x_1^2x_6 + \frac{9}{2} x_2x_6 + 2x_1x_7] \frac{\partial}{\partial x_6}$$
$$+ [\frac{3}{2} x_4 + \frac{1}{2} x_1x_5 + \frac{1}{4} x_1^2x_7 + \frac{7}{2} x_2x_7 + x_1x_8] \frac{\partial}{\partial x_7}$$
$$+ [\frac{1}{2} x_3 + x_1x_4 + \frac{1}{4} x_1^2x_8 + \frac{5}{2} x_2x_8] \frac{\partial}{\partial x_8}\}$$

$$\tilde{\tilde{\mathfrak{X}}}_6 = x_2 \frac{\partial}{\partial x_1} - \{\frac{1}{2} x_1x_3 \frac{\partial}{\partial x_3} + \frac{1}{2} [x_3 + x_1x_4] \frac{\partial}{\partial x_4}$$
$$+ [x_4 + \frac{1}{2} x_1x_5] \frac{\partial}{\partial x_5}$$
$$+ [\frac{1}{2} x_1x_6 + 2x_7] \frac{\partial}{\partial x_6} + [\frac{1}{2} x_5 + \frac{1}{2} x_1x_7 + x_8] \frac{\partial}{\partial x_7}$$
$$+ [x_4 + \frac{1}{2} x_1x_8] \frac{\partial}{\partial x_8}\}. \qquad (2.15\text{-}7)$$

The extended infinitesimal operators have the same commutator table as the unextended ones. Thus they have the same algebraic properties. We are now ready to find the most general class of second order partial differential equations of the form (2.15-3) for each member of the chain (2.15-1).

In general (2.15-3) is invariant with respect to $\mathfrak{X}_i$ iff

$$\tilde{\tilde{\mathfrak{X}}}_i\{x_8 - H(x_1,x_2,\ldots,x_7)\} = 0$$
$$\text{whenever} \quad x_8 = H(x_1,x_2,\ldots,x_7). \qquad (2.15\text{-}8)$$

We consider each member of the chain (2.15-1) in turn and find the corresponding class of partial differential equations invariant under it.

(i) $\{\mathfrak{X}_1\}$.

The condition (2.15-8) implies that

$$\frac{\partial H}{\partial x_1} = 0. \tag{2.15-9}$$

Hence, the most general partial differential equation of the form (2.15-3) invariant under $\{\mathfrak{X}_1\}$ is

$$x_8 = \mathcal{F}(x_2,x_3,\ldots,x_7) \tag{2.15-10}$$

where $\mathcal{F}$ is an arbitrary function of $\{x_2,x_3,\ldots,x_7\}$.

(ii) $\{\mathfrak{X}_1, \mathfrak{X}_2\}$.

(2.15-8) $\Longrightarrow$

$$\frac{\partial \mathcal{F}}{\partial x_2} = 0. \tag{2.15-11}$$

Hence the most general partial differential equation invariant under $\{\mathfrak{X}_1, \mathfrak{X}_2\}$ is

$$x_8 = \mathcal{G}(x_3,x_4,x_5,x_6,x_7) \tag{2.15-12}$$

where $\mathcal{G}$ is an arbitrary function of $\{x_3,x_4,\ldots,x_7\}$.

(iii) $\{\mathfrak{X}_1, \mathfrak{X}_2, \mathfrak{X}_3\}$

(2.15-8) $\Longrightarrow$

$$\mathcal{G} = \sum_{j=3}^{7} x_j \frac{\partial \mathcal{G}}{\partial x_j} . \tag{2.15-13}$$

Solving (2.15-13) we find that the most general partial differential equation invariant under $\{\mathfrak{X}_1, \mathfrak{X}_2, \mathfrak{X}_3\}$ can be expressed in the form

$$x_8 = x_5 \mathscr{A}\left(\frac{x_4}{x_3}, \frac{x_5}{x_3}, \frac{x_6}{x_3}, \frac{x_7}{x_3}\right) \tag{2.15-14}$$

where $\mathscr{A}$ is an arbitrary function of its arguments. Let $\lambda_1 = \frac{x_4}{x_3}$, $\lambda_2 = \frac{x_5}{x_3}$, $\lambda_3 = \frac{x_6}{x_3}$, $\lambda_4 = \frac{x_7}{x_3}$. Then (2.15-14) can be written in the form

$$x_8 = x_5 \mathscr{A}(\lambda_1, \lambda_2, \lambda_3, \lambda_4). \tag{2.15-15}$$

(iv) $\{\mathfrak{X}_2, \mathfrak{X}_6\}$.

The invariance condition (2.15-8) implies that $\mathscr{A}$ must satisfy the partial differential equation

$$\tilde{\tilde{\mathfrak{X}}}_6\{x_8 - x_5\mathscr{A}\} = 0 \quad \text{where} \quad x_8 = x_5\mathscr{A}. \tag{2.15-16}$$

After considerable manipulation (2.15-16) reduces to the following partial differential equation for $\mathscr{A}$:

$$2\lambda_1(1 - \mathscr{A}) = \lambda_2 \frac{\partial\mathscr{A}}{\partial\lambda_1} + 2\lambda_1\lambda_2 \frac{\partial\mathscr{A}}{\partial\lambda_2} + 4\lambda_2\lambda_4 \frac{\partial\mathscr{A}}{\partial\lambda_3} + \lambda_2^2(2\mathscr{A} + 1) \frac{\partial\mathscr{A}}{\partial\lambda_4}. \tag{2.15-17}$$

The corresponding characteristic differential equations are:

$$\frac{d\mathscr{A}}{1 - \mathscr{A}} = \frac{d\lambda_1}{\frac{\lambda_2}{2\lambda_1}} = \frac{d\lambda_2}{\lambda_2} = \frac{d\lambda_3}{\frac{2\lambda_2\lambda_4}{\lambda_1}} = \frac{d\lambda_4}{\frac{\lambda_2^2(\mathscr{A} + \frac{1}{2})}{\lambda_1}}. \tag{2.15-18}$$

The invariants generated from the first and second equalities of (2.15-18) are easily found to be

$$\mu_1 = \lambda_2 - \lambda_1^2 \quad \text{and} \quad \mu_4 = (\mathscr{A} - 1)\lambda_2. \tag{2.15-19}$$

Substituting the expressions (2.15-19) into

$$\frac{d\lambda_1}{\frac{\lambda_2}{2\lambda_1}} = \frac{d\lambda_4}{\frac{\lambda_2^2(\mathscr{A} + \frac{1}{2})}{\lambda_1}}$$

$$\Longrightarrow \qquad \frac{d\lambda_4}{\lambda_2(\mathscr{A} + \frac{1}{2})} = 2d\lambda_1$$

$$\Longrightarrow \qquad \frac{d\lambda_4}{(\mu_1 + \lambda_1^2)\left(\frac{3}{2} + \frac{\mu_4}{\mu_1 + \lambda_1^2}\right)} = 2d\lambda_1$$

$$\Longrightarrow \qquad d\lambda_4 = [3(\mu_1 + \lambda_1^2) + 2\mu_4]d\lambda_1.$$

Hence the third invariant is

$$\mu_2 = \lambda_4 - (3\mu_1 + 2\mu_4)\lambda_1 - \lambda_1^3 = \lambda_4 - \lambda_2\lambda_1 + 2\lambda_1^3 - 2\mathscr{A}\lambda_1\lambda_2. \qquad (2.15\text{-}20)$$

Substituting (2.15-19,20) into

$$\frac{d\lambda_3}{\frac{2\lambda_2\lambda_4}{\lambda_1}} = \frac{d\lambda_1}{\frac{\lambda_2}{2\lambda_1}}$$

$$\Longrightarrow \qquad \frac{d\lambda_3}{\lambda_4} = 4d\lambda_1$$

$$\Longrightarrow \qquad d\lambda_3 = [(12\mu_1 + 8\mu_4)\lambda_1 + 4\lambda_1^3 + 4\mu_2]d\lambda_1.$$

Hence the fourth and final invariant is

$$\mu_3 = \lambda_3 + 2\lambda_1^2\lambda_2 - 3\lambda_1^4 - 4\lambda_1\lambda_4 - 12\mathscr{A}\lambda_2\lambda_1^2 . \qquad (2.15\text{-}21)$$

Hence the general solution of the characteristic differential equations (2.15-18) is

$$\mathscr{F}(\mu_1,\mu_2,\mu_3,\mu_4) = 0 \qquad (2.15\text{-}22)$$

where $\mathscr{F}$ is an arbitrary function of μ_1,μ_2,μ_3 and μ_4. In order to solve for $\mathscr{A}$ there are three major subclasses of solutions of

(2.15-22) of interest.[2]

Case I $\mu_4 = \mathscr{F}_1(\mu_1,\mu_2,\mu_3)$

$$\leftrightarrow \quad x_8 = x_5 + x_3\mathscr{F}_1(\mu_1,\mu_2,\mu_3). \tag{2.15-23}$$

μ_2 and μ_3 depend on $\mathscr{A} = 1 + \dfrac{\mathscr{F}_1}{\lambda_2}$.

An interesting subcase is

$$x_8 = x_5 + x_3\mathscr{F}_1^*(\mu_1) \tag{2.15-24}$$

where $\mathscr{F}_1^*$ is an arbitrary function of μ_1.

Case II $\mu_2 = \mathscr{F}_2(\mu_1,\mu_3,\mu_4)$

$$\leftrightarrow \quad x_8 = \frac{1}{2}\frac{x_7x_3}{x_4} - \frac{1}{2}x_5 + \frac{x_4^2}{x_3} - \frac{1}{2}\frac{x_3^2}{x_4}\mathscr{F}_2. \tag{2.15-25}$$

Here μ_3 and μ_4 depend on

$$\mathscr{A} = \frac{\lambda_4}{2\lambda_1\lambda_2} - \frac{1}{2} + \frac{\lambda_1^2}{\lambda_2} - \frac{\mathscr{F}_2}{2\lambda_1\lambda_2}.$$

An interesting subcase is

$$x_8 = \frac{1}{2}\frac{x_7x_3}{x_4} - \frac{1}{2}x_5 + \frac{x_4^2}{x_3} - \frac{1}{2}\frac{x_3^2}{x_4}\mathscr{F}_2^*(\mu_1) \tag{2.15-26}$$

where $\mathscr{F}_2^*$ is an arbitrary function of μ_1.

Case III $\mu_3 = \mathscr{F}_3(\mu_1,\mu_2,\mu_4)$

$$\leftrightarrow \quad x_8 = \frac{1}{12}\frac{x_6x_3^2}{x_4^2} + \frac{1}{6}x_5 - \frac{1}{4}\frac{x_4^2}{x_3} - \frac{1}{3}\frac{x_7x_3}{x_4} - \frac{1}{12}\frac{x_3^3}{x_4^2}\mathscr{F}_3. \tag{2.15-27}$$

---

[2] The generated functions $\mathscr{F}_1$, $\mathscr{F}_2$, and $\mathscr{F}_3$ are arbitrary differentiable functions of their respective arguments.

Here μ_2 and μ_4 depend on

$$\mathscr{A} = \frac{\lambda_3}{12\lambda_2\lambda_1^2} + \frac{1}{6} - \frac{1}{4}\frac{\lambda_1^2}{\lambda_2} - \frac{1}{3}\frac{\lambda_4}{\lambda_1\lambda_2} - \frac{\mathscr{F}_3}{12\lambda_2\lambda_1^2} .$$

An interesting subcase is

$$x_8 = \frac{1}{12}\frac{x_6 x_3^2}{x_4^2} + \frac{1}{6}x_5 - \frac{1}{4}\frac{x_4^2}{x_3} - \frac{1}{3}\frac{x_7 x_3}{x_4} - \frac{1}{12}\frac{x_3^3}{x_4^2}\mathscr{F}_3^*(\mu_1) \qquad (2.15\text{-}28)$$

where $\mathscr{F}_3^*$ is an arbitrary function of μ_1.

(v) $\{\mathfrak{X}_2, \mathfrak{X}_4, \mathfrak{X}_6\}$

Here we will only consider Case I of (iv). The invariance condition (2.15-8) implies that $\mathscr{F}_1$ must satisfy

$$\tilde{\tilde{\mathfrak{X}}}_4\{x_8 - x_5 - x_3\mathscr{F}_1\} = 0. \qquad (2.15\text{-}29)$$

(2.15-29) reduces to

$$2\mathscr{F}_1 = 2\mu_1\frac{\partial\mathscr{F}_1}{\partial\mu_1} + 4\mu_2\frac{\partial\mathscr{F}_1}{\partial\mu_2} + 3\mu_3\frac{\partial\mathscr{F}_1}{\partial\mu_3} . \qquad (2.15\text{-}30)$$

Letting $\nu_1 = \frac{\mu_2}{\mu_1^2}$, $\nu_2 = \frac{\mu_3^2}{\mu_1^3}$, we find that

$$\mathscr{F}_1 = \mu_1\mathscr{J}_1(\nu_1,\nu_2) \qquad (2.15\text{-}31)$$

where $\mathscr{J}_1$ is an arbitrary differentiable function of ν_1 and ν_2. Hence the corresponding invariant partial differential equation is

$$x_8 = x_5 + x_3\mu_1\mathscr{J}_1(\nu_1,\nu_2). \qquad (2.15\text{-}32)$$

Problem 2.15-1

Corresponding to Cases II and III of (iv) find the most general partial differential equation of the form (2.15-3) invariant under the group generated by the infinitesimal transformations $\{\mathfrak{X}_2, \mathfrak{X}_4, \mathfrak{X}_6\}$.

(vi) $\{\mathfrak{X}_1, \mathfrak{X}_2, \mathfrak{X}_5\}$, <u>the full group of the heat equation</u>.

To find the most general class of partial differential equations invariant under the full group (2.7-14) of the heat equation it is easier to apply $\mathfrak{X}_5$ to the most general partial differential equation (2.15-12) invariant under the subgroup corresponding to the generators $\{\mathfrak{X}_1, \mathfrak{X}_2\}$. The invariance condition (2.15-8) leads to $\mathcal{G}(x_3,x_4,x_5,x_6,x_7)$ satisfying

$$\tilde{\tilde{\mathfrak{X}}}_5\{x_8 - \mathcal{G}\} = 0. \tag{2.15-33}$$

Substituting $\mathcal{G}$ for x_8 in (2.15-33), we find that $\mathcal{G}$ satisfies the partial differential equation

$$\begin{aligned} 2x_3 + 4x_1x_4 + x_1^2\mathcal{G} + 10x_2\mathcal{G} &= (x_1^2x_3 + 2x_2x_3)\frac{\partial\mathcal{G}}{\partial x_3} \\ &+ (2x_1x_3 + x_1^2x_4 + 6x_2x_4)\frac{\partial\mathcal{G}}{\partial x_4} + (2x_3 + 4x_1x_4 + x_1^2x_5 + 10x_2x_5)\frac{\partial\mathcal{G}}{\partial x_5} \\ &+ (12x_5 + x_1^2x_6 + 18x_2x_6 + 8x_1x_7)\frac{\partial\mathcal{G}}{\partial x_6} \\ &+ (6x_4 + 2x_1x_5 + 4x_1\mathcal{G} + x_1^2x_7 + 14x_2x_7)\frac{\partial\mathcal{G}}{\partial x_7}\ . \end{aligned} \tag{2.15-34}$$

(2.15-34) must hold for <u>any</u> values of x_1 and x_2 since $\mathcal{G}$ does not depend on x_1 and x_2. Hence treating (2.15-34) as a polynomial in x_1 and x_2, the coefficients of x_1^2, x_2, x_1 and x_1^o must be set equal to zero. Hence $\mathcal{G}$ must satisfy simultaneously the four partial differential equations:

$$\mathcal{G} = \sum_{j=3}^{7} x_j \frac{\partial\mathcal{G}}{\partial x_j} \tag{2.15-35}$$

$$5\mathcal{G} = x_3\frac{\partial\mathcal{G}}{\partial x_3} + 3x_4\frac{\partial\mathcal{G}}{\partial x_4} + 5x_5\frac{\partial\mathcal{G}}{\partial x_5} + 9x_6\frac{\partial\mathcal{G}}{\partial x_6} + 7x_7\frac{\partial\mathcal{G}}{\partial x_7} \tag{2.15-36}$$

$$x_3 = x_3 \frac{\partial \mathcal{G}}{\partial x_5} + 6x_5 \frac{\partial \mathcal{G}}{\partial x_6} + 3x_4 \frac{\partial \mathcal{G}}{\partial x_7} \tag{2.15-37}$$

$$2x_4 = x_3 \frac{\partial \mathcal{G}}{\partial x_4} + 2x_4 \frac{\partial \mathcal{G}}{\partial x_5} + 4x_7 \frac{\partial \mathcal{G}}{\partial x_6} + (x_5 + 2\mathcal{G}) \frac{\partial \mathcal{G}}{\partial x_7} . \tag{2.15-38}$$

Solving (2.15-35,36) we find that $\mathcal{G}$ reduces to

$$\mathcal{G} = x_5 \mathcal{A}(\mu_1,\mu_2,\mu_3) \tag{2.15-39}$$

where $\mu_1 = \frac{x_5 x_3}{x_4^2}$, $\mu_2 = \frac{x_6 x_3^3}{x_4^4}$, $\mu_3 = \frac{x_7 x_3^2}{x_4^3}$ and $\mathcal{A}$ is an arbitrary function of its arguments.

(2.15-37) requires that $\mathcal{A}$ satisfy the equation

$$1 - \mathcal{A} = \mu_1 \frac{\partial \mathcal{A}}{\partial \mu_1} + 6\mu_1^2 \frac{\partial \mathcal{A}}{\partial \mu_2} + 3\mu_1 \frac{\partial \mathcal{A}}{\partial \mu_3} \tag{2.15-40}$$

and (2.15-38) leads to $\mathcal{A}$ satisfying

$$1 - \mathcal{A}' = (\mu_1 - \mu_1^2) \frac{\partial \mathcal{A}}{\partial \mu_1} + 2\mu_1(\mu_3 - \mu_1) \frac{\partial \mathcal{A}}{\partial \mu_2}$$
$$+ [(\tfrac{1}{2} + \mathcal{A})\mu_1^2 - \tfrac{3}{2}\mu_3\mu_1] \frac{\partial \mathcal{A}}{\partial \mu_3} . \tag{2.15-41}$$

Solving (2.15-40) we find that

$$\mathcal{A} = 1 + \frac{\mathcal{F}(\alpha,\beta)}{\mu_1} \tag{2.15-42}$$

where

$$\alpha = \mu_3 - 3\mu_1, \quad \beta = \mu_2 - 3\mu_1^2$$

and $\mathcal{F}$ is an arbitrary differentiable function of α and β. Substitution of (2.15-42) into (2.15-41) leads to

$$6\mu_1^2 \frac{\partial \mathcal{F}}{\partial \beta} - 2\mu_1 \frac{\partial \mathcal{F}}{\partial \beta} + \mathcal{F} - 3 \frac{\partial \mathcal{F}}{\partial \alpha}$$

$$+ 2\alpha \frac{\partial \mathcal{F}}{\partial \beta} + (\mathcal{F} - \tfrac{3}{2}\alpha) \frac{\partial \mathcal{F}}{\partial \alpha} = 0. \tag{2.15-43}$$

Since μ_1 is independent of α and β, (2.15-43) is a polynomial in μ_1 whose coefficients are independent of μ_1. Hence $\frac{\partial \mathcal{F}}{\partial \beta} = 0$ and $\mathcal{F}(\alpha)$ satisfies the ordinary differential equation

$$(\mathcal{F} - \frac{3}{2}\alpha - 3)\frac{d\mathcal{F}}{d\alpha} + \mathcal{F} = 0. \qquad (2.15\text{-}44)$$

Problem 2.15-2

Find a one-parameter Lie group leaving invariant (2.15-44) and hence show that its general solution is

$$\frac{\mathcal{F}^3}{(2\mathcal{F} - 2 - \alpha)^2} = \text{const.} \qquad (2.15\text{-}45)$$

Hence the most general partial differential equation of the form (2.15-3) invariant under the group of the heat equation is

$$u_{xx} = u_t + \frac{u_x^2}{u}\mathcal{F}(\alpha) \qquad (2.15\text{-}46)$$

where

$$\alpha = \frac{u^2 u_{xt}}{u_x^3} - \frac{3uu_t}{u_x^2}$$

and $\mathcal{F}(\alpha)$ satisfies (2.15-45). In particular, the heat equation corresponds to the case where the const. = 0 in (2.15-45). Show that if the const. $\neq 0$ then the right hand side of (2.15-46) is not a polynomial in the derivatives of u. Hence, the heat equation is the *only* polynomial partial differential equation of the form (2.15-3) invariant under the group of the heat equation.

Problem 2.15-3

For each subalgebra of the chain (2.15-2) find the most general partial differential equation invariant under it.

An Example Where the Transformation of the Independent Variables Depends on the Dependent Variable

For all examples considered so far the Lie group of

transformations of the independent variables did not depend on the dependent variable. We now consider such an example. Our aim will be to find partial differential equations of the form (2.15-3) invariant under the one-parameter Lie group of transformations with infinitesimals

$$\xi(x,t,u) = u, \quad \tau(x,t,u) = 1, \quad \eta(x,t,u) = 0. \tag{2.15-47}$$

Problem 2.15-4

(i) For the infinitesimals (2.15-47) show that the extended infinitesimals are:

$$\eta_x = -u_x^2, \quad \eta_t = -u_x u_t, \quad \eta_{xx} = -3u_{xx}u_x,$$
$$\eta_{tt} = -(u_{tt}u_x + 2u_{xt}u_t), \quad \eta_{xt} = -(2u_{xt}u_x + u_{xx}u_t). \tag{2.15-48}$$

(ii) Show that (2.15-47) leaves invariant the class of partial differential equations of the form

$$u_{xx} = u_x^3 \mathscr{F}(\alpha_1,\alpha_2,\alpha_3,\alpha_4) \tag{2.15-49}$$

where

$$\alpha_1 = u, \quad \alpha_2 = x - ut, \quad \alpha_3 = \frac{1}{u_x} - t, \quad \alpha_4 = \frac{u_t}{u_x} \tag{2.15-50}$$

and $\mathscr{F}$ is an arbitrary function of $\{\alpha_1,\alpha_2,\alpha_3,\alpha_4\}$.

In particular a special case of (2.15-49) is

$$u_{xx} = u_x^2 u_t \tag{2.15-51}$$

corresponding to $\mathscr{F}(\alpha_1,\alpha_2,\alpha_3,\alpha_4) = \alpha_4$.

Problem 2.15-5

Consider the partial differential equation (2.15-51).

(i) Show that the Lie group of transformations leaving

invariant (2.15-51) is of the form

$$\begin{aligned} x^* &= x + \varepsilon\xi(x,t,u) + O(\varepsilon^2) \\ t^* &= t + \varepsilon\tau(x,t,u) + O(\varepsilon^2) \\ u^* &= u + \varepsilon\eta(x,t,u) + O(\varepsilon^2) \end{aligned} \tag{2.15-52}$$

with

$$\begin{aligned} \xi(x,t,u) &= -\gamma x\left[\frac{u^2}{4} + \frac{t}{2}\right] - \frac{\delta}{2} xu + \lambda x + g(u,t) \\ \tau(x,t,u) &= \alpha + 2\beta t + \gamma t^2 \\ \eta(x,t,u) &= \kappa + \delta t + \beta u + \gamma tu \end{aligned} \tag{2.15-53}$$

where $\{\alpha,\beta,\gamma,\delta,\kappa,\lambda\}$ are arbitrary constants and $g(u,t)$ satisfies

$$g_{uu} = g_t.$$

(ii) Note that by interchanging the roles of u and x in (2.15-54) we obtain the group of the heat equation. Hence relate this to the well-known result that if

$$u = \Theta(x,t) \tag{2.15-54}$$

is a solution of (2.15-51) then solving (2.15-54) explicitly for x to obtain

$$x = f(u,t) \tag{2.15-55}$$

we find that f satisfies the heat equation, namely

$$f_{uu} = f_t. \tag{2.15-56}$$

The methods of this section may be applied to the problem of determining general transformations relating the solutions of various partial differential equations. The previous example is the simplest type of such a problem.

APPENDIX

SOLUTION OF QUASILINEAR FIRST-ORDER PARTIAL DIFFERENTIAL EQUATIONS

In Part 2 of this book the invariance of a partial differential equation under a one-parameter Lie group of transformations leads to the consideration of an invariant surface condition for the construction of a similarity (invariant) solution. In the case of two independent variables (x,t) the invariant surface condition corresponds to a quasilinear first order partial differential equation of the form

$$\xi(x,t,u) \frac{\partial u}{\partial x} + \tau(x,t,u) \frac{\partial u}{\partial t} = \eta(x,t,u) \tag{A-1}$$

where $\{\xi,\tau,\eta\}$ are given functions (the infinitesimals of the Lie group) of $\{x,t,u\}$. The solution of (A-1) for the unknown $u(x,t)$ leads to a functional form for u. We now briefly discuss how to solve formally equations of the form (A-1).

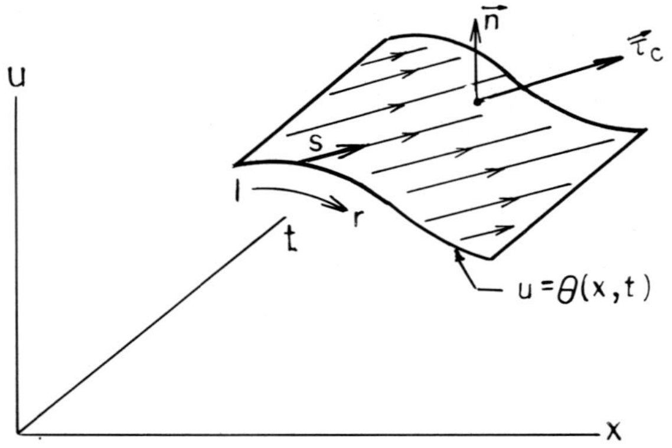

Figure A-1.

---

(1) R. Courant and D. Hilbert, Methods of Mathematical Physics, Vol. II, Interscience, 1962, Chapter II.

Consider, at first, an arbitrary surface in space given by

$$u = \theta(x,t). \tag{A-2}$$

This will later be identified as a solution surface connected to (A-1). We can write the condition that a curve in (u,x,t) space lies in the surface $u = \theta(x,t)$ (see Fig. A-1) as follows:

Let s = parameter along the curve in space so that

$$x = x(s),\ t = t(s),\ u = u(s) \quad \text{defines the curve.} \tag{A-3}$$

Then, the tangent vector to this curve is given by

$$\vec{\tau}_c = \vec{i}_x \frac{dx}{ds} + \vec{i}_t \frac{dt}{ds} + \vec{i}_u \frac{du}{ds}, \quad \vec{i}_x, \vec{i}_t, \vec{i}_u = \text{unit vectors.} \tag{A-4}$$

The normal direction to the surface $u = \theta(x,t)$ can be written easily if we imagine the surface to lie in the family of surfaces

$$S(x,t,u) = u - \theta(x,t) = \text{const.} \tag{A-5}$$

Then the unit normal to $S = 0$ is given by

$$\vec{n} = \frac{\nabla S}{|\nabla S|} = \frac{1}{|\nabla S|}\{-\vec{i}_x\theta_x - \vec{i}_t\theta_t + \vec{i}_u\}. \tag{A-6}$$

The condition that the curve $x(s),\ t(s),\ u(s)$ lie in the surface is thus $\vec{n}\cdot\vec{\tau}_c = 0$ or

$$\theta_x \frac{dx}{ds} + \theta_t \frac{dt}{ds} = \frac{du}{ds}. \tag{A-7}$$

Thus, if we define a family of curves (characteristics) locally by the differential equations

$$\left.\begin{aligned} \frac{dx}{ds} &= \xi(x,t,u) \\ \frac{dt}{ds} &= \tau(x,t,u) \\ \frac{du}{ds} &= \eta(x,t,u) \end{aligned}\right\} \text{(characteristic differential equations)} \tag{A-8}$$

then (i) each one-parameter family of such characteristic curves generates a surface which is an integral surface $(u = \theta(x,t))$ of

$$\xi(x,t,u)\frac{\partial u}{\partial x} + \tau(x,t,u)\frac{\partial u}{\partial t} = \eta(x,t,u) \qquad \text{(A-9)}$$

and

(ii) conversely, each such integral surface is generated by a one-parameter family of characteristic curves.

The family of solutions of the characteristic differential equations can be represented parametrically by

$$\begin{aligned} x &= x(s,r) \quad \text{where} \quad s = \text{parameter along a characteristic curve,} \\ t &= t(s,r) \quad r = \text{parameter identifying a characteristic} \\ u &= u(s,r) \quad \text{curve} = \text{const. on a characteristic.} \end{aligned} \qquad \text{(A-10)}$$

The condition that the parameters (r,s) can be eliminated so that we can obtain a solution surface, one value of u for each (x,t), is the condition that the Jacobian

$$J = \frac{\partial(x,t)}{\partial(s,r)} = \frac{\partial x}{\partial s}\frac{\partial t}{\partial r} - \frac{\partial t}{\partial s}\frac{\partial x}{\partial r} \neq 0. \qquad \text{(A-11)}$$

For our applications we are interested in functional forms or in the so-called general solution of the basic equation (A-1), containing one arbitrary function. One way to approach the general solution is to consider the general initial value problem, that is, to look for the solution surface passing through an arbitrary initial curve given by

$$s = 0 \quad \text{and} \quad x = x_o(r), \quad t = t_o(r), \quad u = u_o(r). \qquad \text{(A-12)}$$

We assume that the initial curve has nowhere the characteristic direction so that

$$J(0,r) = \xi(0,r)\,\frac{\partial t_o}{\partial r} - \tau(0,r)\,\frac{\partial x_o}{\partial r} \neq 0. \tag{A-13}$$

Then a unique solution exists, at least in the neighborhood of the initial curve. Further, it can be shown that for the linear case $\{\xi(x,t),\ \tau(x,t)\}$, $J(s,r) \neq 0$. However, for the non-linear case the solution may fail to exist when $J(s,r) = 0$. This can, for example, correspond to a folding or turning back of the solution surface in (x,t,u). Such cases have not arisen in the problems considered here.

Example (i). Find the general solution of

$$xu_x + tu_t = u. \tag{A-14}$$

The characteristic differential equations are

$$\frac{dx}{ds} = x,\ \frac{dt}{ds} = t,\ \frac{du}{ds} = u. \tag{A-15}$$

The one-parameter family of integral curves passing through $x_o(r)$, $t_o(r)$, $u_o(r)$ at $s = 0$ is thus

$$x(s,r) = x_o(r)e^s,\ t(s,r) = t_o(r)e^s,\ u(s,r) = u_o(r)e^s. \tag{A-16}$$

The parameters may be eliminated by

$$\frac{x}{t} = \frac{x_o(r)}{t_o(r)},\quad e^s = t.$$

Since r = const. on a characteristic, $\frac{x}{t}$ = const. on the projections of the characteristics on the plane $u = 0$. r can be eliminated in favor of (x,t). A simple way is to choose $x_o(r) = r$, $t_o(r) = 1$. Then

$$r = \frac{x}{t},\quad e^s = t \tag{A-17}$$

and

$$u = tu_o\left(\frac{x}{t}\right),\quad \text{the general solution,} \tag{A-18}$$

where $u_o\left(\frac{x}{t}\right)$ is an arbitrary (differentiable) function.

A short-hand version of the idea above is used in the calculations of this book. The characteristic differential equations are written free of parameters as

$$\frac{dx}{x} = \frac{dt}{t} = \frac{du}{u} . \tag{A-19}$$

Integration of the first two of (A-19) defines the projections of the characteristic curves on (x,t)

$$\frac{x}{t} = \text{const.} = r. \tag{A-20}$$

Then, the variation of u along the characteristics can be found by integrating either

$$\frac{dx}{x} = \frac{du}{u} \quad \text{or} \quad \frac{dt}{t} = \frac{du}{u} \tag{A-21}$$

along $r = \text{const.}$, remembering that the "constant" of integration is thus an arbitrary function of r. For example,

$$\log t = \log u - \log u_o(r) \tag{A-22}$$

or

$$u(x,t) = tu_o\left(\frac{x}{t}\right) \tag{A-23}$$

as before. (Note that the form obtained by using the other equation $u(x,t) = xu_o^*\left(\frac{x}{t}\right)$ is equivalent.)

Exactly the same idea can be used for any case with linear left hand side. It is sufficient to consider

$$\frac{dx}{\xi(x,t)} = \frac{dt}{\tau(x,t)} = \frac{du}{\eta(x,t,u)} \tag{A-24}$$

and to follow the procedure above.

For the general non-linear case, however, the system of equations

$$\frac{dx}{\xi(x,t,u)} = \frac{dt}{\tau(x,t,u)} = \frac{du}{\eta(x,t,u)} \tag{A-25}$$

must be solved simultaneously. The solution may take an implicit form.

Example (ii). Find the form of the general solution of

$$uu_x + u_t = t. \tag{A-26}$$

The characteristic differential equations are

$$\frac{dx}{u} = \frac{dt}{1} = \frac{du}{t} . \tag{A-27}$$

Integrating the last two shows that

$$u - \frac{t^2}{2} = \text{const.} = r \text{ (say) on a characteristic.} \tag{A-28}$$

Thus, the first two become

$$\frac{dx}{r + \frac{t^2}{2}} = \frac{dt}{1} \tag{A-29}$$

which integrates to

$$x = rt + \frac{t^3}{6} + F(r) \tag{A-30}$$

where F is an arbitrary (differentiable) function of r. (A-30) may be written

$$u = \frac{x}{t} + \frac{t^2}{3} - \frac{1}{t} F(u - \frac{t^2}{2}) \tag{A-31}$$

or equivalently

$$u = \frac{t^2}{2} + G(ut - x - \frac{t^3}{3}). \tag{A-32}$$

In this case the general solution can only be obtained in implicit form. This result is connected with the fact that the solution to the initial value problem may cease to exist when $J = 0$ as s departs from its initial value.

BIBLIOGRAPHY

PART 1

1. A. Cohen, An Introduction to the Lie Theory of One-parameter Groups with Applications to the Solution of Differential Equations, Boston, 1911.

2. L. E. Dickson, Differential equations from the group standpoint, Annals of Math., 25, 1924, pp. 287-378.

3. L. P. Eisenhart, Continuous Groups of Transformation, Dover, 1961.

4. K. O. Friedrichs, Lectures on Advanced Ordinary Differential Equations, Gordon and Breach, 1965, Chapter I.

5. E. L. Ince, Ordinary Differential Equations, Dover, 1956.

6. S. Lie, Gesammelte Abhandlungen, Leipzig and Oslo, 1922 (reprinted 1960).

7. S. Lie, Theorie der Transformationsgruppen, Vol. I (1888), Vol. II(1890), Vol. III(1893), Leipzig, (reprinted by Chelsea Publishing Company, New York, 1970).

8. S. Lie, Vorlesungen über Continuierliche Gruppen mit geometrischen und anderen Anwendungen, Leipzig, 1893 (reprinted by Chelsea Publishing Company, New York, 1967).

9. S. Lie, Vorlesungen über Differentialgleichungen mit bekannten infinitesimalen Transformationen, Leipzig, 1891 (reprinted by Chelsea Publishing Company, New York, 1967).

10. S. Lie, Verhandlungen der Gesellschaft der Wissenschaften zu Christiania, November, 1874.

11. J. M. Page, Ordinary Differential Equations with an Introduction to Lie's Theory of the Group of One Parameter, Macmillan, 1897.

12. H. Weyl, On the Simplest Differential Equations of Boundary Layer Theory, Ann. of Math., 43, 2, 1942, pp. 381-407.

BIBLIOGRAPHY

PART 2

1. S. Lie, Über die Integration durch bestimmte Integrale von einer Klasse linearer partieller Differentialgleichungen, Arch. for Math., Vol. VI, No. 3, Kristiana, 1881, p. 328.

2. L. V. Ovsjannikov, Gruppovye Svoystva Differentsialny Uravneni, Novosibirsk, 1962. (Group Properties of Differential Equations, translated by G. Bluman, 1967).

3. E. A. Müller and K. Matschat, Über das Auffinden von Ähnlichkeitslösungen partieller Differentialgleichungssysteme unter Benutzung von Transformationsgruppen, mit Änwendungen auf Probleme der Strömungsphysik, Miszellaneen der Angewandten Mechanik, Berlin, 1962, p. 190.

4. P. W. Bridgman, Dimensional Analysis, 2nd Ed., Yale University Press, 1931.

5. L. I. Sedov, Similarity and Dimensional Methods, Moscow, 6th Ed. (in Russian); English translation, 4th Ed., Academic Press, (1959).

6. G. Birkhoff, Hydrodynamics, 2nd Ed., Princeton University Press, 1960.

7. G. I. Barenblatt and Ya. B. Zel'dovich, Self-similar solutions are intermediate asymptotics, Ann. Rev. of Fluid Mech., 1972.

8. G. W. Bluman, Construction of Solutions to Partial Differential Equations by the Use of Transformation Groups, Ph.D. thesis, California Institute of Technology, 1967.

9. G. W. Bluman and J. D. Cole, The general similarity solution of the heat equation, J. of Math., and Mech., Vol. 18, No. 11, May, 1969, pp. 1025-1042.

10. G. W. Bluman, Applications of the general similarity solution of the heat equation to boundary-value problems, Quart. of Appl. Math., Vol. 31, No. 4, January 1974, pp. 403-415.

11. G. W. Bluman, Similarity solutions of the one-dimensional Fokker-Planck equation, Int. J. Non-lin. Mech., 6, pp. 143-153, 1971.

12. L. L. Ovsjannikov, Gruppovye Svoystva Uravnenya Nelinaynoy Teploprovodnosty, Dok. Akad. Nauk, CCCP, 1959, 125, 3, p. 492.

13. W. F. Ames, Nonlinear Partial Differential Equations in Engineering, Academic Press, 1972, Volume II, Chapter 2.

14. A. J. A. Morgan, The reduction by one of the number of independent variables in some systems of partial differential equations, Quart. J. of Math., 3, Ser. 2, 1952, pp. 250-259.

15. A. G. Hansen, Similarity Analyses of Boundary Value Problems in Engineering, Prentice-Hall, 1964.

16. G. Rosen and G. W. Ullrich, Invariance group of the equation $\frac{\partial u}{\partial t} = -\underline{u}\cdot\nabla\underline{u}$, SIAM J. on A. Ma., May, 1973, Vol. 24, No. 3, pp. 286-288.

16. J. D. Cole and J. Aroesty, Hypersonic similarity solutions for airfoils supporting exponential shock waves, A.I.A.A. Journal, 8, February, 1970, No. 2, pp. 308-315.

18. P. Germain, Écoulements Transsoniques Homogènes, Progress in Aeronautical Sciences, Vol. 5, Pergamon Press, 1964, pp. 143-273.

INDEX